Safety, Health and Environmental Auditing

A Practical Guide, Second Edition

Safety, Health and Environmental Auditing

A Practical Guide, Second Edition

By

Simon W. Pain

CRC Press
Taylor & Francis Group
Boca Raton London New York

CRC Press is an imprint of the
Taylor & Francis Group, an **informa** business

CRC Press
Taylor & Francis Group
6000 Broken Sound Parkway NW, Suite 300
Boca Raton, FL 33487-2742

Printed on acid-free paper

International Standard Book Number-13: 978-1-138-55715-4 (Hardback)

Library of Congress Cataloging-in-Publication Data

Names: Pain, Simon Watson, author.
Title: Safety, health, and environmental auditing : a practical guide / Simon W. Pain.
Description: Second edition. | Boca Raton : Taylor & Francis, CRC Press, 2018. | Includes bibliographical references and index.
Identifiers: LCCN 2017046275| ISBN 9781138557154 (hardback : alk. paper) | ISBN 9781315150673 (ebook)
Subjects: LCSH: Environmental auditing. | Industrial hygiene--Auditing. | Industrial safety--Auditing.
Classification: LCC TD194.7 .P35 2018 | DDC 658.4/08--dc23
LC record available at https://lccn.loc.gov/2017046275

Visit the Taylor & Francis Web site at
http://www.taylorandfrancis.com

and the CRC Press Web site at
http://www.crcpress.com

For Mark, Catriona, Fraser and Jack.

The future's safe in your hands.

Contents

Preface

This new edition of *Safety, Health and Environmental Auditing* builds on the success of the earlier book. It has been enhanced to embrace new topics, including 'due diligence' EHS auditing and process safety auditing, and a new chapter summarises the relevant international standards on auditing. The previous material has also been updated to align with the guidance and requirements set out in the ISO standards.

Most people would agree that health and safety is important. Those who consider environmental protection also to be important are a probably smaller but rapidly growing number. Unfortunately, that is often where all interest in these subjects ends. It is all too easy to say what should have happened after there has been some adverse event such as an accident, injury or environmental release, but why can we not be wise enough to recognise these shortcomings before things go wrong and therefore avoid hurting either people or the environment?

Many competent organisations have extensive safety, health and environmental instructions in place but still find that things often go wrong. The problem is one of human behaviour. People like to make life easy for themselves and therefore sometimes ignore the instructions, or perhaps the instructions themselves are out of date. In the latter part of the last century, it was realised that this was the cause of production 'quality' problems, and so the quality improvement processes which culminated in such ISO standards as 9001, 14001 and 45001 were introduced. It was realised that having good quality instructions was not enough. What really mattered was how well people adhered to those procedures. A crucial part of a good quality process is the checking (or auditing) step to ensure that people are complying with the procedures.

It was quickly realised that a similar checking process could be of great benefit with respect to compliance with safety, health and environmental management procedures. In the 1980s, some leading companies started to carry out environmental audits and later on branched into health and safety. The results of these audits were dramatic and often resulted in as much as a 10-fold reduction in incident frequency rates. Consultants quickly realised that there was a demand to be satisfied in helping organisations improve their environmental health and safety performance and started to provide high-quality auditing services. The consequence for the organisation was that they achieved a significant improvement in their performance, but it sometimes came with a rather large price tag in the form of consultants' fees.

The purpose of this little book is to provide 'down to earth' guidance for managers and specialists in those organisations who are committed to improving their safety, health and environmental performance, but either are not sure where to start or cannot or do not wish to employ consultants to do this for them.

The book is intended for those managers and safety/environmental specialists who have some level of safety, health and environmental awareness. It has been written in such a way that it is easy to dip in and out of the short chapters to refresh your memory, prior to or during an audit. A set of audit protocols covering 60 different aspects of environmental, health and safety management are provided in Appendix 2, for those who have not developed their own. An electronic copy of these protocols is available for download from the Solway Consulting Group website (www.solway-consulting.com) to allow for easy copying and printing for the audit. Frequently needed practical administrative checklists which may be useful when planning and conducting the audit are found in Appendix I.

For those who prefer an all-electronic audit checklist, a copy of the Plaudit 2 audit protocol is also available for download from the Solway Consulting Group website. This allows the auditor to complete his or her notes in real time and provides a continuous graphical audit compliance score. It must be remembered that the electronic protocol is merely a supporting tool and is no substitute for a detailed understanding of how to prepare for and how to conduct the audit.

Good luck!

About the Author

Simon W. Pain is an award-winning independent safety, health and environmental management consultant based in Scotland. He has a wealth of health, safety and environmental management experience in various manufacturing industries gained over the last 45 years.

Simon is a chartered mechanical engineer with more than 32 years' experience in senior management positions with British Steel, ICI and DuPont's engineering, manufacturing, research and corporate functions. He has been advising company executives at the board level on safety, health and management issues for the last 20 years. He spent many years as divisional safety, health and environmental manager for ICI and DuPont; the latter is widely regarded as the world benchmark company for health and safety standards. During the last 20 years, he has developed novel techniques in health and safety training and communication that were commended by the Institute of Occupational Safety and Health in November 2004. He has subsequently developed the unique SHEEMS emergency management system for small businesses. It is for this work that he has been awarded Best International Health and Safety Consultancy in Scotland and also International Health and Safety Consultancy of the Year 2017.

As a consultant, Simon specialises in raising awareness and motivating senior managers to achieve a paradigm shift in health and safety awareness. He does this by using the high-impact approach and making the subject interesting and fun.

He is an expert in auditing, especially at the management level and personally designed and developed the ICI audit protocol system to ensure that auditing standards were consistent. As a Det Norske Veritas–trained auditor he has led audits not only in the United Kingdom but also in the United States, Japan, India and throughout Europe. He regularly carries out lectures and training for health, safety and environmental auditing.

He is a fellow of both the UK Institute of Mechanical Engineers and the Institute of Energy and a chartered member of the Institute of Occupational Safety and Health. He was also a member of the UK government's Energy Best Practice Committee and a member of the Solway River Purification Board until the formation of the new Scottish Environment Protection Agency.

Simon graduated in mechanical engineering from the University of Birmingham and obtained his postgraduate qualifications in health and safety from the University of Loughborough.

More information may be found at www.solwayconsulting.com.

1 Elements of a Good Safety, Health and Environmental System

A 'system' is 'an environment exploiting, restricting and repressing individuals'. So claims the *Collins Concise Dictionary*. Surely this cannot be the intention of safety, occupational health and environmental systems? Perhaps a more appropriate definition would be 'a way of doing things'. However, *Collins* is right in suggesting that systems may not necessarily be a help; they can on occasion be a hindrance. We have all experienced the uniformed official who insists on rigidly applying outdated rules with the claim that 'it is more than his job's worth not to comply!' Nevertheless, systems are needed in organisations, whether they cover the control of finances, the payment of employees, the purchase of goods, the control of product quality or the application of safety, health and environmental (SHE) standards. Although we may sometimes doubt it, systems are created to simplify activities that are repeated and are essential to the purpose of the organisation. They are intended to ensure that we benefit from the learning and experience of others, so that we do not all have to go back and reinvent the activity from first principles. Even when formal systems do not exist, it is human nature for us to want to make things easy for ourselves, so we often tend to devise our own way of doing things.

The role of the system in collating experience and learning is an essential component when systems are intended to prevent harm occurring to people or the environment around us. This is why the application of systems to SHE protection is of such importance and explains the recent explosion in regulatory controls in this area from governments around the world. Indeed, governmental controls are now so complex that new systems have to be introduced to try and simplify the previous systems. It is hardly surprising that Collins defines these systems as restricting and repressing individuals.

The problem with any system is that it tends to start to deteriorate from the first day it is introduced. This can be through ignorance, oversight or wilful disregard. Ignorance is fundamentally a communication and training issue, and wilful disregard is arguably a disciplinary matter, but probably the biggest barrier to the successful application of systems in the SHE area is oversight. Oversight, or the inability to anticipate adverse consequences, is one of the most common causes of harm to people and their surroundings. So often it is done with the best of intentions. No one intentionally crashes his or her car into a brick wall. It may happen because of a desire to get home on time. That is a creditable enough intention, but all too often the best intention ends in tragedy because people don't think about the consequences of driving around a corner too fast, ignoring the driving safety control systems of speed limits or failing to make allowances for tyre condition or adverse weather.

Systems exist in all walks of life, and there is ample evidence that they deteriorate with time unless they are properly managed and controlled. The people who are to operate the system need to know what is expected of them. The requirement should be written down in a clear and concise manner and then this should be effectively communicated to the individuals concerned through training. The system should appear to be sensible and logical to the people involved. If a system appears to be illogical, then there will be a natural tendency for people to devise an alternative approach that they consider to be more appropriate. Ideally, the system should be developed by one or more of the people who will have to operate it and certainly not by some remote bureaucrat who never leaves the sheltered cloisters of his or her own office. When people have been trained in the system, their knowledge of the requirement should then be checked to ensure a thorough understanding before they are asked to apply it. Records should be maintained to confirm that users have been trained in the system and to identify when refresher training will be required. Even then, with a sensible logical system and trained operatives, the survival of the system is not assured unless it is properly managed. Most systems fail quite quickly, usually within 6 to 9 months, unless people are reminded of the need (i.e. an accident happens and reinforces the need for a safety system) or management ensures compliance with the system through some process of checking or auditing.

Regrettably, the effort required by management to ensure compliance to SHE management systems is not always seen as a top priority when more pressing problems arise. Often it is only after some accident or injury occurs that the investigating team goes back to the safety instructions and finds to its relief that the injured party had transgressed some detailed sub-clause in the dusty document. Management feels vindicated and the poor wretch is given a summary dressing down to add to his or her physical ailments. Perhaps if a little more thought were given, the management team might recognise that it had also failed in its task of ensuring compliance with the safety procedure. Managers are well advised to remember that they have a 'duty of care' towards their employees and that the regulatory authorities are likely to want to know as part of their investigation exactly what steps management took to ensure compliance with both regulatory and its own internal safety procedures. The mere existence of a written procedure does not confirm that there was necessarily an effective system in place. In fact, the existence of an injury almost immediately suggests that management has failed in its duty of care. It is now well understood by experienced incident investigators that the immediate causes of an incident are rarely the real underlying cause of the event. Frequently, the immediate causes relate to the period in the few minutes prior to the incident. The underlying causes go back much farther in time and often have roots in management's lack of control over a prolonged period and the failure to have robust SHE systems in place.

The consequences of failures in safety are often all too immediate because often someone gets hurt, but system failures in protecting employees' health from exposure to asbestos dust may not be realised for 40 years, and failures in environmental controls to prevent land contamination may not result in observable consequences for even longer. Compliance with such environmental and occupational health protection systems is particularly difficult as responsible parties may consider that the risk of the consequences coming back to haunt them within their career span is so small as to be worth taking a chance.

Leaders in the field of industrial loss prevention all advocate for the same three elements of a safe and healthy working environment. These I shall refer to as the ‘three Ps’ and are ‘people’, ‘procedures’ and ‘plant’. Taking these in reverse order, let us consider ‘plant’ first. Plant or operating equipment needs to be of adequate standard in order to achieve a good safety and health record for the workforce and its neighbours. Work equipment should be properly designed to be safe to use and should have been subject to an appropriate form of risk assessment. Often, plant and equipment has the potential to deteriorate with age, and so the standards of maintenance and upkeep are critical. ‘Procedures’ should be established to ensure that potentially hazardous equipment remains safe to use and in a condition that will not cause harm to the environment. These critical elements of plant and equipment maintenance are known as SHE assurance and are an important part of corporate governance. However perfectly designed and maintained plant and equipment may be, it cannot alone ensure that no harm occurs. Equipment is used and operated by people, and people are notoriously unreliable. The use of equipment should be controlled by the second of the three Ps: namely, ‘procedures’. In the late 1980s, when the International Quality Standard series ISO 9000 was first being implemented on a large scale, many companies made the mistake of believing that everything could be controlled by procedures and instructions. Every eventuality was considered and the tropical rainforests disappeared overnight in a mountain of procedural bibles. The main problem was that the sheer volume of procedures was unmanageable, and they were rarely used and never revised. It is now recognised that you cannot ‘proceduralise’ every aspect of life and that the procedures should relate to the important and generic activities. The application of good and well-maintained procedures allows a step change improvement in SHE performance, compared with relying only on a well-designed and well-maintained plant.

To achieve world-class performance in SHE management, we require the involvement not just of engineers, designers and managers but also the proactive involvement of all employees. Employees must become responsible for not just their own safety but also that of their workmates; they should in effect become ‘their brother’s keeper’. Experience in hazard recognition training demonstrates to me that people will identify more hazards when pooling their ideas and working as a team than any one individual will do when working alone. Consequently, to have an effective loss prevention system requires attention to the plant, the procedures and the people.

The scope of a loss prevention audit may cover any combination of safety and occupational health and environmental protection. It is logical to attempt to combine these three issues as they all relate to harm to individuals, groups of individuals or the environment. They are all issues of ‘loss prevention’. Safety harm, in the form of injuries, usually arises as a result of acute effects and is often short term and reversible (except in the case of fatalities), whereas occupational illness usually relates to long-term exposure and results in chronic effects. However, the underlying causes for both sets of consequences can be the same. Audits often need to pay particular attention to health hazards or environmental effects because the consequences are not immediate and therefore may be less obvious to the worker. Unfortunately, because of the size of the task, the danger of attempting to cover all aspects of health, safety and environmental control within a single audit is that the audit either becomes unwieldy or at the other extreme may become superficial in its individual elements. This problem of superficiality is the greatest practical problem facing the health and safety auditor today. Superficiality not only discredits the outcome of an individual audit but may bestow a feeling of inappropriate 'comfort' when this may not be fully justified.

2 Management Systems

According to estimates by the International Labor Office (ILO), the number of job-related accidents and illnesses annually claims more than 2.3 million lives worldwide, and this number appears to be rising because of rapid industrialisation in some developing countries. The assessment also indicates that the risk of occupational illness has become by far the most prevalent danger faced by people at work—accounting for 2 million annual work-related deaths and outpacing fatal accidents by nearly six to one.

The ILO found that in addition to work-related deaths, each year there are some 268 million nonfatal workplace accidents in which the victims miss at least 3 days of work as well as 160 million new cases of work-related illness. To put this into perspective, this is like saying that every man, woman and child in the United States will have a work-related injury every year! Injuries, illnesses and environmental incidents are costly not only to the world's economy, but also to workers, their families and our surroundings.

In many countries, company directors, managers and employees can now be held personally liable for failure to control health and safety. Increasingly, managers are held criminally liable when things go wrong, and so there is an increasing tendency for organisations to document their safety systems. Of course, a mere written procedure does nothing in itself to reduce the risk of harm to employees; it is merely a statement of intent. To translate such a statement into meaningful action requires some sort of management activity. To ensure that this action is properly sustained requires monitoring by the management team. The level of informality or formality of the system will depend on the nature of the enterprise and the risks associated with it. The essential starting point is to consider SHE management as a key business process. The board or senior management of the organisation should set down its basic requirements in the areas of SHE protection in the form of a policy statement, which should be made available to all employees. The policy should state the organisation's position on SHE matters and how all the employees will be expected to comply with them. It should also state the arrangements and responsibilities within the organisation for implementation of that policy. The policy should influence all the organisation's activities, including the selection of people, equipment and materials, the way work is done and how goods are designed and services are provided.

In summary, the policy should

- Be a clear written statement of the organisation's position relating to loss control in safety, health and the environment
- Identify who is responsible for SHE performance
- Identify the sources of expert SHE knowledge
- Be signed by the most senior person(s) in the organisation
- Be prominently displayed in an up-to-date form
- Be communicated in clear and concise terms to everyone within the organisation

The existence of an up-to-date policy statement is a clear indication that the management team considers safety, health and environmental loss prevention to be a key issue for the organisation. However, as stated previously, the existence of such a statement does not avoid accidents happening. To make the policy effective, it is necessary to get the employees involved and committed. Creating positive loss prevention behaviour among the staff needs to be properly managed. First, people need to know how they are expected to behave in the organisation, what tasks they are required to do, and how, where and when they should do them. It is the responsibility of management to set the standards of behaviour that are required, with a view to controlling the risks to employees, customers, neighbours and the environment. Many industry standards already exist, and in some cases, it is appropriate to adopt these. In other cases, it will be necessary for the organisation to develop its own standards or requirements. Either way, the standards should identify the basic management requirements for loss prevention, but they must be documented, measurable, achievable and realistic if they are to be effectively adopted.

Organisations wishing to develop their own standards should consider these areas of their operations for application of those standards:

- Premises and workplace
- Assets design and procurement
- Substance control and material hazards
- Transport and distribution
- Storage and warehousing
- Task design and risk assessment (safe systems of work)
- Training requirements
- Continuous improvement plans
- Product safety
- Regulatory compliance
- Change control
- Construction
- Maintenance
- Environmental control
- Health assessments
- Emergency and crisis management
- Contractor management
- Effluent and wastes
- Office and laboratory safety
- Energy and water conservation
- Spillage prevention and control
- Atmospheric emission abatement

Frequently, standards will be stated in a general 'high level' way which is either not 'user-friendly' or covers a wider scope than particular employees may require. Very often, standards will state 'what' has to be done but not necessarily 'how' it should be achieved or 'who' is responsible for doing it. In these circumstances, it may be necessary to develop further guidance for the employees. Guidance usually takes the form of a record of best practice. Good examples of the provision of guidance are in the UK government Environmental Technology and Energy Efficiency Best Practice Programmes and the US Occupational Safety and Health Administration (OSHA) and UK Health and Safety Executive (HSE) guidance notes. These programmes do not mandate how the user is to save energy or define what environmental technology to procure; rather, they provide information on the best way of approaching the problem. The user is then left with choices regarding which solution to adopt. It is very much the same with the provision of guidance in your management system—think of your guidance documents or records as part of the memory of your organisation. If you think it sounds like a lot of effort, then just try running an organisation with amnesia.

But standards and guidance are largely management information documents; they don't help Charlie to produce gizmos from his high-speed press nor do they help Daphne understand what to do to minimise the risk of work-related upper-limb disorder when using her computer or tablet. People at the sharp end of the organisation need clear job instructions because so often they are the ones most likely to be at risk of injury or occupational illness. It is very common when we have that new machine which doesn't work that we resort to the instructions only when all else has failed. In a SHE-conscious culture, we want behaviours to apply so that complying with the operating instructions is the normal and accepted way of life. To be valued and useful, the instructions should be clear, concise and unambiguous. Ideally, the people who use them should write them and the instructions should be regularly reviewed to ensure that they represent the current best practice. If someone comes up with a better way of doing things, then this should not be adopted except under controlled trial conditions until the instruction has been changed. As soon as one permits operators to deviate from the instruction, then it encourages employees to believe that deviating from other instructions is also acceptable.

An effective instruction should identify the following:

- The purpose of the instruction
- The scope (What circumstances does it apply to and what does it not apply to?)
- Technical term definitions
- Relevant cross-references (Keep these to a minimum as it complicates the process of updating instructions later on if instructions refer to one another.)
- Who is responsible for carrying out the task
- What records need to be kept
- The procedure to be followed

Having identified the organisation's SHE policy, set standards and produced working instructions, organisations very often heave a sigh of relief in the belief that their system is in place. Unfortunately for them, the most difficult part of establishing the system is still to come. Without gaining the commitment of staff and training the staff in the application of the instructions, all your valuable documented standards and instructions will lie on the shelf gathering dust. The key element in any system, which is so often overlooked, is that of 'implementation'.

Managers now have to do what they are paid for: they must implement the system. That means organising and motivating their staff to apply the instructions and to recognise opportunities for improvement. Managers are responsible for employing competent staff, so their first task is to ensure that they recruit staff appropriate for the job. The next step on the road to establishing full competence is effective training—something that we say so easily but find so difficult to deliver. After each step of training is complete, then it is necessary to validate the learning and confirm that the employee has achieved the required level of understanding. Of course, being trained is very different from being competent. The final step in developing competence is to put the new skill safely into practice. During this early application of a new skill, the new trainee should be 'mentored' by an experienced person in the subject who can keep a check to ensure that the trainee is applying the newly acquired skill in the correct manner. So often, we train people in the right skills but at the wrong time. It must be remembered that 'practice makes perfect' and lack of practice leads to the skill being lost or eroded. Even when our employees are fully competent, we cannot relax. We need to keep them updated through regular communication on matters of safety, health and the environment so that they are aware of new hazards, risks and preventative measures.

The next step in the SHE management system is to monitor the performance of our employees and the organisation, to ensure that people are doing what we expect them to do. This checking process, which has a similar purpose to that carried out by financial auditors, is the task we know as 'SHE auditing'. This periodic checking process will identify issues that are not correct. Sometimes, the issues will be minor, but on other occasions, they will be more significant. These issues were in the past normally referred to as 'noncompliances' or 'nonconformances'. Each noncompliance represents a potential hazard that could lead to an undesirable consequence. Hazards, once identified, cannot go unchecked, and so actions have to be devised to prevent future repetition of the nonconformance. These events are called 'corrective actions'. The process of monitoring performance through auditing, leading to the identification of nonconformances, which in turn leads to corrective action, is the essence of an ongoing continuous improvement process. The continuous improvement process is a never-ending upward spiral of learning. It has no point at which it is complete, as it is an ongoing process, not a programme with a beginning and an end. If this process is rigidly adhered to, although minor noncompliances will occur, the major collapse of safe systems, which is so often apparent after the investigations into some of the world's major industrial incidents, can be avoided.

Some large companies operating from a large number of locations have adopted the financial 'letter of assurance' concept in relation to their SHE management processes. In this way, the corporate headquarters requires each operating unit to provide an annual letter indicating how that part of the organisation complies with the company standards and instructions. This enables the company board members to discharge their responsibilities against its policy but also has the little-recognised consequence that personal liability gets passed down through the organisation. So remember that if your boss asks you for a letter of SHE assurance, he or she is actually handing over some of his or her personal liabilities to you.

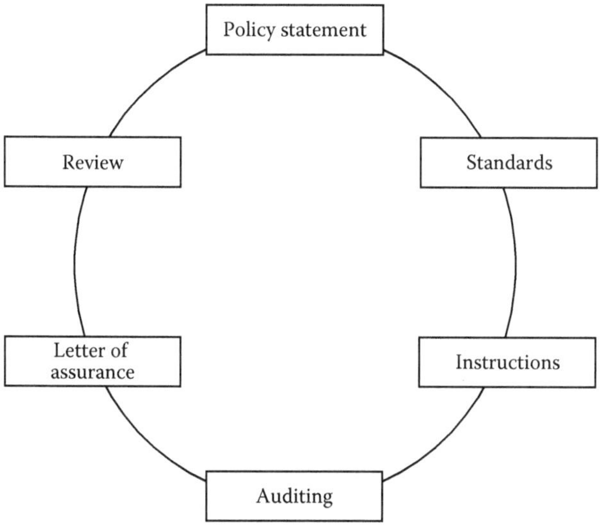

FIGURE 2.1 The SHE management process.

Any effective system or procedure has a review step built into it, to ensure that it stays up to date and benefits from the learning that arises from application. The final step in a good SHE management system is no different. Every year, the senior management team of the organisation should review its SHE policy and standards in the light of the experience of the previous year and consider whether any changes are appropriate.

To recap, the SHE management process is summarised in Figure 2.1.

A good SHE management system starts with setting a policy. The basic management requirements to achieve this policy are then established as the organisation standards. These standards are then interpreted into useful and practical instructions for personnel to use at the operating level. Management then arranges for the instructions to be implemented, ensuring that the relevant people are trained and competent to carry out their tasks in a safe and healthy manner without damage to the environment. The system is then subject to periodic checking or auditing to ensure that the local instructions are complied with. Any noncompliances are then subject to corrective action. The senior management team then reviews the full system, once a year, to ensure that any learning from the application of the process is built into the future policy or standards. In accountancy audits, there is a requirement for the organisation's management to prepare a letter of assurance confirming that the management teams have taken steps to comply with accounting procedures. These letters of assurance are now starting to be required by the boards of directors of some organisations where they require departments or operating divisions to confirm compliance with SHE standards. In these circumstances, the best method of being able to confirm regulatory or standards compliance in the letter of assurance is via a robust auditing programme.

3 Auditing
The Principles

Contrary to the view of many business pundits, the overriding principle of business is not about maximising profit but about avoiding loss. Organisations can survive without making a profit but soon cease to operate if they consistently make a loss. The management of SHE issues is all about the prevention of loss, be that the loss of life, the loss of health, the loss of environmental heritage, the waste of energy or the loss of time and other scarce resources. Historically, the management of safety in particular has centred on analysing the records of injuries – which in effect means waiting until the management system breaks down and results in someone getting hurt and then responding by putting a number of fixes in place to deal with the recommendations from the investigation. The quality management approach that has been successfully adopted to overcome the same 'When it's broke, fix it' approach in the manufacturing situation is progressively being used in the safety and environmental areas to great effect. Initially, the ISO 9000 range of general quality standards was used to apply to environmental protection, but now separate standards exist in the form of international standards ISO 14000, ISO 19011 and the ECO Management and Audit Scheme (EMAS) for ensuring environmental management system compliance and OHSAS 18001 for safety and occupational health compliance (soon to become ISO 45001). The key change that the quality approach has brought to loss prevention activities is the element of audit and auditability. This single concept potentially moves our systems and procedures from something that is short-lived and changes whenever the boss moves on, to a situation when it not only has longevity but also progressive improvement.

The word 'audit' is derived from the Latin 'auditus', which means 'a hearing'. Until fairly recent years, the practice of auditing was largely limited to assessing the reliability of company financial accounts, with the intention of ensuring that good accountancy practice was being followed and to root out any financial irregularities. The practice of it being a 'hearing' soon moved from the friendly tête-à-tête to something more akin to a judicial hearing. Financial auditors these days require high levels of investigative skill and are not averse to being judgemental in their reporting. The modern principle of a SHE audit is that it is a regular, systematic check of the system in order to assess whether the organisation or working group's performance meets the required performance. ISO 9001 defines quality audits as follows:

> Systematic and independent examinations to determine whether quality activities and related results comply with planned arrangements and whether these arrangements are implemented effectively and are suitable to achieve objectives.

The same definition could be adopted for SHE auditing by replacing the word 'quality' with 'safety, health and environmental'. However it is defined, the audit should attempt to establish:

1. The level of understanding of the standards or requirements
2. The degree of conformance with those requirements
3. The adequacy of the requirements
4. The steps necessary to achieve further improvement
5. The extent to which regulatory requirements are met

In addition, especially in the case of ISO 14000 and EMAS environmental audits, the purpose may also be to permit the organisation to be accredited and authorised to display that accreditation to customers and suppliers.

Audits should normally be conducted on some periodic frequency according to a predetermined audit plan, but they may also be prompted by some significant change or event in the organisation. A change of management, a particular incident or the need to follow up on a particular noncompliance may prompt the need for an unscheduled audit. The frequency of scheduled audits is a judgement for the local management team. It will be a balance between the time required for the audits themselves and the time required to implement corrective actions. Some organisations have experienced the problem of 'audit fatigue' when formal audits become so frequent that they start being resented or corrective action requests are ignored. In these circumstances, the audit process becomes discredited and worthless. However, once an audit plan is established, adherence to the plan will be seen by members of the organisation's staff as a measure of senior management's commitment to safety, health and environmental matters in general and not just to the audit process.

Safety, health and environmental auditing varies significantly from the quality management auditing required by such standards as ISO 19000 and ISO 10011. Accredited quality management auditing by the well-known national and international accreditation bodies usually assesses the auditee only against its own procedures because of the multitude of requirements of all the different types of enterprises. Although ISO 9000 provides a useful framework, it does not and never can provide examples of best practice for every eventuality. This, of course, raises the question not so much about whether some of the less reputable organisations are complying with their own procedures, but rather, whether those procedures are good enough. ISO 9000 also suffers from the problem of being rather unwieldy for small and medium-sized enterprises, and although the Quality Guild provided by the local training and enterprise companies (TECs) attempts to fill this gap, it has so far met with limited success.

Safety auditing, in particular, is different. There are some generic standards that apply to almost every workplace. Almost every organisation has at some time experienced injuries from slips, trips or falls. It is possible to assess the risk to personnel of injury from these causes in almost any situation. Consequently, the experienced safety auditor can provide not only the observation of the potential for tripping hazard, but he or she can also proffer advice on solutions. Similarly, in occupational health auditing, it does not matter whether the organisation is a school classroom or a steel works, an auditor can make useful observations regarding whether the noise levels are likely to cause harm to the occupants and whether protective measures are adequate. Unlike many areas of quality management auditing, the SHE auditor can and should make observations on both the compliance with the local standard *and the adequacy of the standard itself.*

Auditing is all about evaluating the performance of an organisation or part of an organisation and comparing that with a standard. Frequently, audits will identify corrective action; in fact, if audits habitually return back the message that everything is OK, as was the case in the routine permit to work auditing on the ill-fated Piper Alpha* drilling platform, then one should start to question the effectiveness of the auditing process. If, however, the only message to come back from the audit process is a large list of corrective actions, there is a danger that the process will be seen as negative and overly critical. One or two corrective actions must not be allowed to overshadow the wealth of good things that the audit identifies are going on. The first objective of an SHE audit should be to *recognise and give credit for the particularly good things that are observed.* Not only is this encouraging to the auditees, but it also provides an excellent training opportunity for the auditors and allows them to network ideas back into other audits or their own workplace. The audit is not just a continuous improvement process for the audited unit but is also a continuous improvement and education process for the auditors.

In larger organisations, it may not be possible to satisfactorily design a single audit that will cover all aspects of SHE management. In fact, in large multinational companies, very large numbers of such audits are carried out every year. Whatever the size of the organisation, the person arranging or requesting the audit must consider the purpose of the audit in order to avoid the problem of superficiality by attempting too broad a scope. The SHE auditor is interested in examining three main areas. He or she wishes to know, first, *what is it that the organisation claims that it should be doing* in regard to SHE management? These questions should be answered by looking at the organisation's policy, standards and instructions. The next task that the auditor needs to assess is *what the organisation should be doing.* In other words, are the standards that the organisation has set itself good enough? How do they compare with best practices or at the very least the minimum regulatory requirements and industry codes of practice? Finally and most importantly, the auditor wants to establish *what really happens* in the organisation. Are the procedures being carried out in the way that management expects or has some form of malpractice started to creep in? It may even happen that both management and employees believe that they are operating the system, but they interpret it in different ways. Some years ago, there was a problem in the United Kingdom when automatic half-barriers were first introduced at railway level crossings. The sign at the roadside proclaimed, 'Do not cross while red lights are flashing.' In parts of Yorkshire, the local dialect uses the word 'while' in the way that many others use the word 'until', and so, although everyone had a clear understanding of what the railway sign meant, some people did not cross while the lights were

* The Piper Alpha disaster in the North Sea in July 1988 was the world's worst offshore oil disaster in terms of fatalities. One hundred sixty-five people died from the fire and explosion that followed a pump being recommissioned with its safety valve removed for maintenance.

flashing, whereas others in Yorkshire in particular waited until the lights flashed to cross! Clearly, a situation existed where everyone thought that he or she knew what the requirement was, but a local anomaly had caused it to be ambiguous.

The nature of larger organisations is that they often have multiple site operations. Consequently, the companies set broad standards at the corporate level, which are interpreted into local procedures at the site or individual facility level. This hierarchical approach results in an interesting and unique approach to auditing, which also has to be tiered. Typically, there are three levels of auditing, as shown in Figure 3.1, although there is no common agreement about the convention for numbering the tiers; some companies count down, whereas others count up. Consequently, talking about 'Level 1' health and safety auditing leads to confusion. To minimise misunderstanding, we refer to the levels as 'management', 'specialist' and 'operational'.

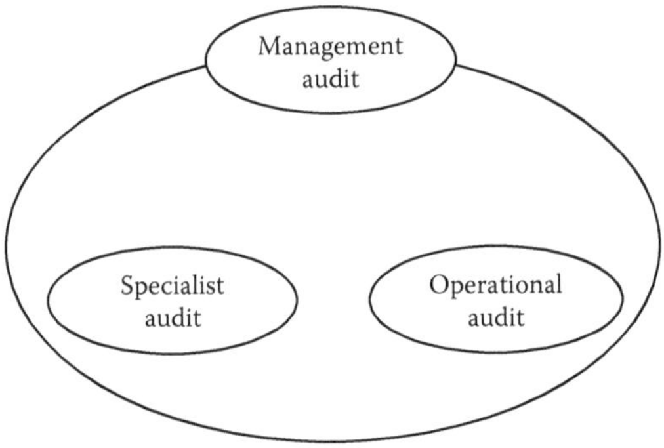

FIGURE 3.1 Audit levels.

Whatever the level of the audit or the complexity of the scope being audited, the process of the audit will be very similar. The process is one of detection. Using the concept of the author Hans Christian Anderson, as auditors, we are trying to establish whether the emperor is wearing those new clothes he has told us about, and are they as good as he would have us believe or is it all a mirage? The auditor needs to appreciate what is required and then discover what actually happens. By comparing the desired state, (i.e. the requirement or standard) with the actual state (i.e. what actually happens), it is possible to identify whether a gap exists between these two states and identify whether any actions are necessary to ensure that the actual state and the desired state converge.

The process followed in most cases will typically be a version of that shown in Figure 3.2.

The standard or requirement is first condensed into a manageable checklist. There is then a process of data gathering which involves talking to people and looking at what happens. The information gleaned via this route then needs to be confirmed through some process of verification before the auditor can draw his or her conclusions and make recommendations for improvement.

This sequence of audit actions is the framework of this book and, through the use of the text and appendices, should enable any capable environmental, safety or occupational health professional to conduct efficient and beneficial audits with the minimum of cost and upset.

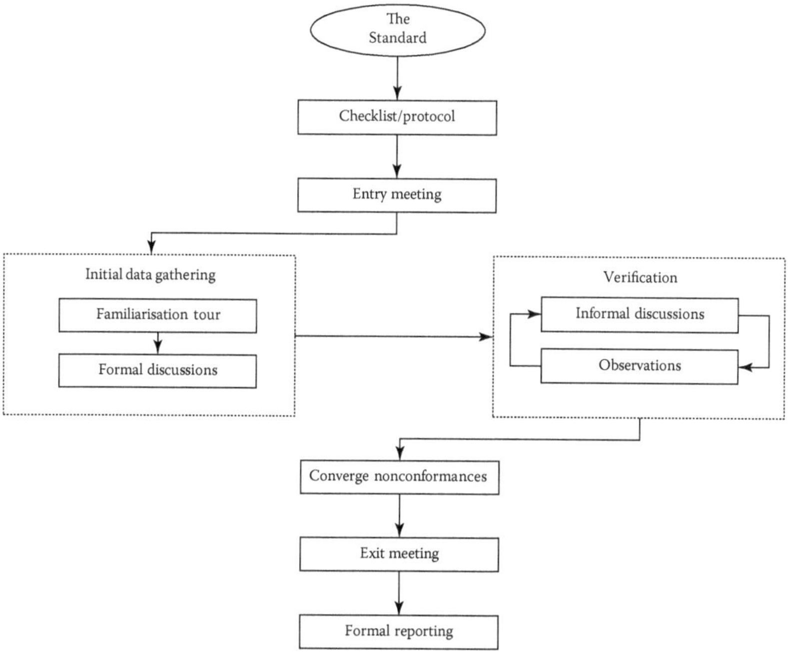

FIGURE 3.2 Typical audit process.

MANAGEMENT AUDITS

Management audits are done at the 'strategic' level in order to examine the adequacy of arrangements for managing SHE affairs in an organisation. They occur at a high level and are intended to monitor whether the management team is aware of all the health and safety issues relevant to its specific operations. In particular, the auditor should assess for each aspect of SHE management whether there is a procedure in place to address that aspect, whether the relevant people are trained and validated in the procedure, and whether the procedure has been subject to periodic review in the light of practical learning and regulatory changes that may have occurred. The auditor will usually be asked to identify a level of compliance with the overarching standard, be that a corporate requirement or a regulatory one. In summary, the management audit is an assessment of whether the facility management is managing all the health and safety issues that it should be doing. A team of senior managers and senior safety professionals from outside the facility management team usually carries out management-level audits.

SPECIALIST AUDITS

We have seen that management audits are intended to cover the full spectrum of health and safety issues relevant to a single site. As a consequence, it is not possible to drill down to any great depth into any subject in an audit that may attempt to cover up to 100 different aspects. The specialist audit goes to the other extreme and examines single topics in great depth. In this case, a single or set of related aspects or topics may be considered by a specialist who is an 'expert' in that area. The topic may be an occupational health issue, such as manual handling, or a safety issue, such as equipment integrity. In the former case, the auditor would perhaps be an occupational physician, and in the latter case, an experienced electrical or mechanical engineer. In the management hierarchy, if management audits are thought of as 'strategic', then specialist audits equate to 'tactical' management.

The role of the specialist auditor is not only to assess whether the corporate and local procedures are being applied but also to assess, in the light of his or her specialist knowledge, whether the local procedures are good enough and take account of recent learning. It includes some measure or validation of the system. The specialist auditor will also do an assessment of the depth of understanding that exists in the particular aspect of safety, health or the environment. In particular, have key personnel changed or is the experience of those people responsible adequate? If specialist engineering audits had been carried out at Nypro* Ltd at Flixborough in the United Kingdom in the early 1970s, they may have uncovered that there was no professionally qualified and experienced engineer on-site and that the workshop staff did not have the experience to appreciate the consequences of designing the now infamous reactor bypass pipe by sketching it in chalk on the workshop floor.

OPERATIONAL AUDITS

Numerically, by far the largest number of audits will be carried out by local managers auditing compliance with their own procedures. These audits, which are similar to the ISO 9000 range of local audits, are usually known as 'operational' or 'compliance' audits and deal with the detail of how things are to be done and what workplace precautions are required. In this case, local managers (or other interested and skilled employees such as safety representatives) are trained in the full range of auditing skills, aimed at understanding how to assess compliance with a health, safety or environmental procedure. It has been common in many industries to apply this approach to safe systems of work issues, but the approach is increasingly being used as a means of auditing the full range of health, safety and environmental issues. The slightly intimidating consequence of applying this approach across the board is that it can result in a large number of audits each year, and it is a common problem for audit schedules to fall behind the plan. However, the benefits hugely outweigh the disadvantages. Effective operational-level auditing not only confirms compliance levels and identifies opportunities for improvement, but it also has great benefits educationally for both the auditor and the auditee. It is always said that 'the best way to learn is to teach'. The auditor cannot effectively audit unless he or she knows the requirements himself. The commercial airline industry is the most effective exponent of operational-level auditing. Everyone recognises that human error by pilots

* Twenty-eight workers were killed when the Nypro (UK) plant at Flixborough exploded in June 1974. Another 36 were injured. A few days prior to the incident, one of the five cyclohexane reactors was removed and replaced by a large bypass pipe. This modified pipework failed, causing a massive vapour cloud explosion.

can potentially lead to large loss of life. The airline industry and their regulators have responded to this by establishing very regular compliance-level auditing of pilots during flights and flight preparation. To carry out thorough monthly checks on all pilots is a huge task and has the potential to lead to audit fatigue. Instead of using independent auditors, it is done by a 'third pilot', who is a regular pilot, but for that flight his role is changed and he becomes an auditor. One day, the third pilot may be auditing a captain, and another day, he himself may be being audited by that same captain. This compliance-level auditing process is now so well established that it is an accepted and valued part of every pilot's work to both be audited and to audit. This also establishes the power of learning through auditing. Effective auditing also immediately identifies whether procedures are out of date, which is often the biggest complaint of employees on the shop floor. When operational auditing begins in earnest, it is the author's experience that initially more than 75% of procedures will need revision and updating. As soon as managers say that a procedure is obviously out of date and employees should know that and 'use their initiative', they must also recognise that they have failed in their duty to provide a safe system of work.

It is worth re-emphasising here the difference between the ISO 9000 quality audit and the full SHE audit. Quality audits comprise two main depths of audit, often known as the 'systems' audit, which examines the quality system with a view to confirming that it follows the quality manual requirements, and the 'compliance' audit, which is an in-depth examination with view to assessing compliance. The SHE audit has great similarities to the compliance audit, but there is no equivalent of the specialist SHE audit in ISO 9000. This step in the SHE audit is a critical one because it assesses whether the standards that are being set are good enough.

The three elements of the SHE auditing are summarised in Figure 3.3 and may require different frequencies or even different auditor knowledge, depending on the size of the operation.

In particular, the second level of audit, sometimes known as the 'specialist audit', may require auditors with particular knowledge of electrical engineering or industrial hygiene, for example. Furthermore, the auditing of the policy and standards may need to be done only every few years, whereas the compliance-level audit, which is checking what actually happens at the sharp end for some local instructions such as safe systems of work, may need to be carried out daily or weekly. For most organisations of medium size, it should be possible to combine the management and specialist requirements in a single audit, but it will always be necessary to keep the compliance audit at a frequency that makes it appropriate for the programme to be

Level 1 audit	'What we actually do'	Compliance audit	Operational
Level 2 audit	'What we should do'	Specialist audit	Tactical
Level 3 audit	'What we say we do'	Management audit	Strategic

FIGURE 3.3 The audit hierarchy: summary of audit levels.

managed and conducted by local staff. The frequency of compliance-level auditing will depend on the aspect being audited. For example, permits to work may need to be audited weekly or monthly, whereas evacuation procedures may need to be done only annually.

PURPOSE AND BENEFITS

In any 'rule-based' society, the mere presence of the rule itself does not usually guarantee that people will comply with that rule. Consequently, means of enforcement are usually developed. In the case of national laws, the enforcement is usually done via the police and the courts of law. An individual who is found failing to comply with the law will usually face some sort of adverse consequence involving a fine or imprisonment.

Similarly, in the case of safety rules within an organisation, the existence of the rule itself will not effect a long-term change in employees' behaviour. It is necessary for management not only to establish the rules or standards by which it expects to operate but also to monitor whether the standards are being applied. When behaviour differs from the standard, then corrective action must be applied.

The principles of establishing an effective long-term SHE management system follow the principles of 'plan', 'do' and 'evaluate'. How this links to the importance of auditing as a key part of the 'evaluate' or monitoring step is shown in Figure 3.4.

It should be noted that auditing relates not just to the 'act' step but also relates to the 'planning and specifying' stage and to the setting of SHE policy. The findings of the auditing or monitoring step are then fed back to the previous step to ensure that there is a continuous improvement process.

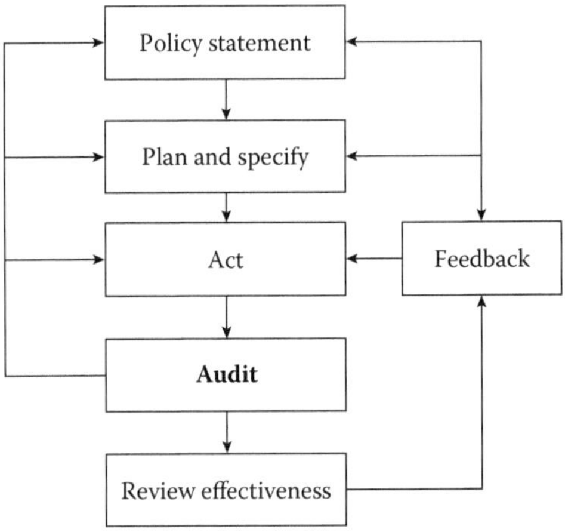

FIGURE 3.4 Auditing as part of the SHE management process.

From this model, it can be seen that auditing is a crucial part of the management system. It is the author's experience that even when procedures or instructions exist, they are often irrelevant documents that sit on remote shelves and are hardly used, and bear little in common with what actually happens in practice. The prime purpose of auditing is to ensure that what should happen (i.e. what is in the procedures) is actually what happens in practice. It is possible to do this in a qualitative or quantitative way, or both. Waring, in his book *Safety Management Systems* (published by Chapman & Hall, London, 1996), identifies the following objectives of the safety audit:

- Validating safety policy and strategy
- Testing safety compliance and verifying progress
- Establishing the current level of safety performance
- Identifying areas of high hazard and risk issues
- Summarising the current strengths and weaknesses
- Producing prioritised action lists and plans
- Setting future safety performance targets
- Improving the management of resources

It is interesting that Waring makes no reference to behavioural modification as one of the objectives. Auditing is by definition a 'human factors' issue, and some of the commercially available proprietary auditing systems focus almost exclusively on behavioural modification.

In the study undertaken by the Accident Prevention Advisory Unit at the Health and Safety Executive (Success and Failures in Accident Prevention), the summarised conclusion states:

> Any simple measurement of performance in terms of accident frequency rate or accident/incident rate is not seen as a reliable guide to the safety performance of an undertaking.

This conclusion is perhaps borne out by the events in Mexico City,* where the organisation had an excellent performance in terms of injury frequency rate, but there was still a catastrophe leading to the loss of many lives. The salutary lesson for managers responsible for the safety of their operations is that SHE management is a multifaceted task and there is a need to manage all aspects of the task to avoid adverse consequences. There is no quick fix, and focusing on one aspect alone brings with it the risk of losing sight of other equally important factors. This shows one of the primary benefits of the management-level audit: it provides a periodic check on whether the balance of SHE management effort is becoming skewed or distorted.

The Accident Advisory Unit report finds there is no clear correlation between such a simple measurement of injury frequency rate and the work conditions, in injury potential or in the severity of injuries that have occurred. A need exists for more accurate measurement so that a better assessment can be made of efforts to control foreseeable risks. It is suggested that more meaningful information would

---

* Fifteen workers were killed following a major explosion in an isomerisation unit at the BP refinery in Texas City in March 2005. The plant was being recommissioned following a major overhaul when excessive hydrocarbon flow to the blowdown system resulted in flammable liquids carrying over out of the top of the flare stack, accumulating on the ground and causing a vapour cloud, which ignited. Most of the victims were maintenance contractors who were still occupying the temporary cabins nearby.

be obtained from the systematic inspection and auditing of physical safeguards, systems of work, rules and procedures, and training methods than from data about accident experience alone.

However, it is the author's experience that hard-nosed business managers are looking for more quantifiable benefits from health and safety audits. In this case, the benefits of audits can be seen to fall into six categories, as shown in Figure 3.5.

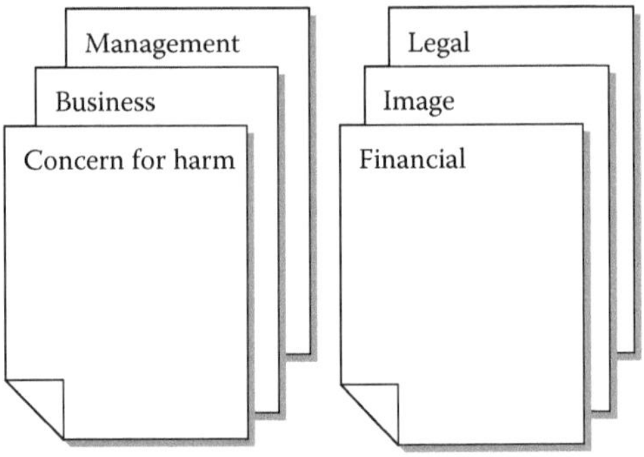

FIGURE 3.5 Benefits of audits.

1. Concern for harm – The caring approach and concern for harm will be demonstrated by fewer injuries and greater employee loyalty.
2. Financial benefits – Avoidance of increased insurance premiums, increased operating costs, fewer process interruptions, all of which may lead to enhanced shareholder value.
3. Business benefits – Fewer business interruptions, less lost time, improved profitability.
4. Image benefits – Personal pride and improved public relations. Enhanced corporate image can lead to improvements in recruitment quality.
5. Management benefits – Include 'peace of mind', confidence in standards, a continuous improvement process and a consistent approach (avoiding repeated safety initiatives).
6. Legal benefits – The avoidance of fines, imprisonment, legal costs and civil claims.

Avoidance of fines is often thought to be a motivator in management behaviour, but in reality, unless the fines are punitive, the level of the fine tends to be relatively low, particularly when compared with the resources and funds of medium- and large-sized organisations. In reality, the cost of litigation and the loss of image are often of much greater consequence. However, the frequency of imprisonment for managers is on the ascendancy, particularly in cases of safety negligence. The reality is that as caring managers, we need to demonstrate to those people who work for us that we actually care about them and want them to go home unharmed to their families at the end of the working day. We want our people to know that 'our work is never so urgent or important that we can't take the time to do it safely'. Carrying out audits is one way of showing people that as managers we really are committed to safety and environmental improvement.

The effect of image damage on organisations was highlighted some years ago, when a large multinational company had a leak from an oil pipeline that passed under a large river. Although the actual amount of environmental damage was small, the publicity arising from the incident resulted in a dearth of graduate recruitment applications for a period of several years. It appeared that new graduates do not want to work for companies that they consider to be environmental polluters.

The British Chemical Industry Safety Council report 'Safe and Sound' states that the top management of the US chemical companies noted for profitability as well as safety were convinced that effective loss prevention programmes were essential for a company's prosperity and accepted as part of good business. The report suggests that one requirement of these programmes was that their effectiveness must be checked by safety audits to ensure that an organisation's assets are safeguarded.

In 1991, research carried out by Dr Larry Gaunt at Georgia State University Center for Risk Management shows that the mere action of auditing usually has a beneficial effect. Gaunt's review of 33 major users of the International Safety Rating Scheme proprietary health and safety audits indicates that more than two-thirds of the auditees reported a positive effect from audit systems.

Gaunt's survey results

Positive effect	68.3%
No effect	29.8%
Negative effect	1.5%
Not applicable	0.4%

Tight adherence to SHE standards will result in fewer accidents and environmental incidents. Investigating these events is disproportionately demanding of management's time and the real costs are often hidden. Research by the UK Health and Safety Executive has shown that these hidden costs can be as much as 36 times the immediately obvious costs.

It is clear from earlier comments that there are significant benefits from SHE audits, but it is also evident from the wide range of audit processes that are currently in use that some audits have different objectives and different levels of success. In general terms, having good SHE standards is about applying good management practice or a 'quality' approach to how we manage our people (health and safety) and the situations in which we live and work (the environment).

4 What Makes a Good Auditor?

Although ISO 19011 ('Guidelines for Quality and/or Environmental Management Systems Auditing') is written primarily for environmental systems audits, it has some useful general advice on how to plan audit programmes. In particular, it identifies five principles of auditing:

1. Ethical conduct
2. Fairness
3. Professional
4. Independence
5. Evidence based

These principles can readily be applied to what makes a good auditor. To behave ethically, the auditor must be trusted and must have integrity and discretion. It is not the auditor's job to go and 'tell tales' about the audit findings to anyone who will listen. The auditor must be fair and ensure that he or she reports the audit in a truthful way that accurately represents the facts as they were discovered. It is essential that the auditor appears to be professional and that he or she is competent to carry out the audit, is diligent and careful and has the ability and experience to be able to apply judgement. This last point is crucial in the auditor's credibility, as a mindless application of 'rules' will not endear the auditor to the auditees. It is also important that the auditor has some level of independence and has no conflict of interest. An audit carried out by the local safety manager may not be independent if his or her annual bonus depends on the outcome of the audit score. Finally, the audit must be based on evidence that is verifiable and will stand up to challenge.

Auditing is all about 'hearing', and so the key competency is that the auditor must be a good listener who has the respect of the people being audited. He or she should be trained in auditing techniques and be personally knowledgeable about the subject of the audit. Above all, the auditor must be open-minded and able to recognise and value different ways of achieving the same end. Criticism that the auditee has not resolved a problem in the same way that the auditor would have done is not likely to win too many friends. The auditor must establish a position of helping the audited unit rather than just leaving it with a group of insoluble problems. I always make a point of finding an opportunity to provide information, contacts or further personal help after the audit is completed, as a way of indicating the principle of the whole process being one of working together, rather than having a confrontational attitude between auditor and auditee.

Selecting the number of auditors for an audit is often difficult to get right. My principle is always to err on the side of less rather than more. There is nothing that irritates the auditees more than finding an army of auditors descending on them and then finding that there is not enough for all the auditors to do. On a major audit that may take several days, it is advisable for the auditors to be seen to be committed and hard working (i.e. 'diligent' as required by ISO 19011), rather than spending all their evenings in the local bar and being seen to be on a corporate jaunt. Compliance-level audits of local procedures rarely need more than one auditor, who would normally be a local employee or manager. Such audits would normally select one particular instruction or subject to audit and would rarely take more than an hour. At the other extreme, a management/specialist audit covering the full range of SHE standards could take up to a week, and OSHA audits in the United States have been known to last as long as 15 weeks, although those are very much the exception. These types of audits would typically have two to four auditors, depending on the size and diversity of the audited unit and whether there were any trainee auditors in attendance. In selecting the auditor team for such a major audit, consideration should be given to ensuring that the auditor team provides the following experience:

- Formal auditing training
- Prior auditing experience
- Experience of similar activity to that carried out in the audited unit
- A thorough understanding of the relevant regulatory requirements
- Sufficient seniority to stand up to the local senior manager
- Knowledge of the local language and culture (if overseas)
- Professional SHE knowledge

A typical thorough SHE management audit will identify a large number of issues, many of which will be *de minimus* and trivial. The auditor must be able to recognise the significant issues and present those in a way that is not prescriptive and leaves some room for management to personalise the solution and hence provide the opportunity for local ownership. It is usually better for the auditor to flag the problem, rather than prescribe the solution. Remember, that we are on a shared journey towards never-ending improvement, and the auditor's primary role is to point auditees in the direction they should be going, as so often local managers have a multiplicity of possible directions in which to proceed. We need to provide guidance to their directionless signs (Figure 4.1). As soon as the auditor ceases to be 'helpful', a large part of the benefit of the audit has been lost.

Auditors should expect to be challenged on their findings and conclusions and should be confident and robust when supporting their arguments. This is why recommendations must be based on verifiable evidence. A recommendation based on little fact will collapse under scrutiny and will not only eliminate a possibly valid point but will also cast doubt on the quality of all the other recommendations and even the competence of the auditor.

Finally, because auditing is all about contact with people and often happens in a situation that can be seen by the auditee as potentially threatening, auditors' own interpersonal skills are critical. So often, a pompous, self-opinionated auditor who arrives full of the powers invested in him or her by the head office will be politely (or otherwise) tolerated and everyone will just want to get rid of the auditor as soon as

FIGURE 4.1 The directionless sign.

possible and get back to doing things the way they were done before. A well-organised, experienced auditor who recognises the location's qualities and wants to work jointly towards a further improvement in SHE performance is much more likely to receive a good reception and to influence future thinking and facilitate further improvement.

It is usually wise to allow the auditor to focus attention on a specific topic. The 'safety performance assessment' concepts used by some companies wishing to carry out mini-specialist audits exemplify this approach. In this case, a particular topic is taken as the subject for the assessment, and the auditor selects a small number of aspects of that topic to assess on a range of applications, which usually attempts to be the majority of applications of that topic being applied on that day. For example, the assessment topic may be 'safe access', 'spillage control' or 'load-bearing/ground stability'. In the case of safe access, the assessor then decides to examine the 'adequacy' of the access to the scaffold or other working platform. The assessor then confirms the legal and technical requirements for scaffolding access, and examines every, or the majority of, scaffolds in his or her jurisdiction against these criteria. Usually, auditing tries to cover a broad range of activities on a sampling basis, in the belief that random samples will be statistically representative of the overall situation. The safety performance assessment type of audit does the reverse, by taking a very narrow subject and attempting to do a 100% examination of the applications.

The skilled auditor is likely to adopt a 'questioning' approach rather than a 'telling' approach. This approach applies the 'push versus pull' theory (expounded in *Use Your Head* by Tony Buzan, published by Pearson, London, 2006). We can think of the auditee as a cup that is either full or empty.

The 'empty cup' theory is that people are devoid of knowledge and need to be sat down and told or 'filled up' with knowledge. This is the traditional approach to education and is characterised by a 'tell' style of communicating. In an audit scenario, this would mean that the auditor talks and the auditee listens. Teaching children to play football for the first time uses the 'empty cup' approach. They need to know the rules before they try to play the game. They have little prior knowledge, and so we, as their parent and coach, tell them what they have to do and which goal to aim for. However, telling a professional footballer or a mature football fan the same information would be considered condescending and might even result in getting a bloodied nose.

On the other hand, the 'full cup' theory uses the idea that auditees are brimming over with ideas and experience, but some of this information is hidden in their subconscious. The function of the auditor is to draw out that experience by the use of probing questions. Frequently, the auditees know what they should be doing and don't need the auditor to tell them. Questions such as *'What could go wrong here?'* or *'What could be done to reduce this risk?'* will invariably get the auditee to identify his or her own corrective actions. The problem is that many auditors are too ready to propose solutions rather than elicit answers. Using an asking approach rather than a telling approach establishes respect for auditees and recognises that they have a wealth of knowledge. So the auditor's approach should normally be one of posing a series of open questions. If, after asking the probing questions, it becomes clear that the auditee doesn't have a clue, then it may be appropriate to proffer some advice or to seek the information elsewhere.

Concern for impact, an enquiring mind and an ability to pose sensible questions are probably the auditor's greatest armoury and should be taken as much more important than technical knowledge when selecting the audit team. But many of the skills that are required to make a good auditor are the same as those required to make a good manager or coach. Good organisational skills are essential as a multi-aspect management audit is an opportunity to create mayhem within the audited organisation. The auditor should arrange audit discussions, tours and site visits to minimise disruption to the auditees. There is nothing worse for an auditee than to be called back to see the auditor time after time to discuss different aspects of the audit, just so that the auditor can follow his preordained audit sequence. In this situation, the auditor should rearrange the aspects being audited into a single audit discussion. In other words, don't expect people to rearrange their commitments to fit in with your disorganised audit programme.

On a major audit that spans several days, the experienced auditor also considers his or her own capacity to absorb information. Several days of nonstop audit discussions are inclined to make the brain hurt, and so the auditor should be capable of sufficient self-discipline to blend the very intense discussion periods with time out on the site. This not only provides a break and some variety for the auditor but also raises the auditor's profile by being seen out and about and talking to people at the sharp end of the business.

5 The Standard or Requirement

One definition of auditing is 'a process of systematic examination to assess the extent of conformances with defined standards and recognised good practice and thereby identify opportunities for improvement'. As has been mentioned before, an audit differs from an inspection primarily because an audit compares what exists with some defined requirement, whereas an inspection often uses what is in the inspector's mind as the benchmark. It all sounds easy enough, but what do we mean by a 'standard' and can we be sure that our standards are the same as those of the organisation or department that we are auditing?

A standard is simply an agreed measure or requirement. For example, the standard for measuring length is either the imperial or the metric system, depending on which side of the Atlantic you reside. By having a recognised standard, such as the mile or the kilometre, we intrinsically understand how far away places are and this allows us to do other things easily, such as assess how long journeys might take or work out the approximate costs of travel. Often the standards for things of such importance as weights and lengths are laid down by august international committees. Standards of national importance are usually laid down by governments and go under the grand title of laws and regulations. It is interesting to note that many of our legal standards are derived from adverse events of the past and are actually quite philanthropic, as their primary intention is to prevent the recurrence of some past harm or disaster.

A common set of legal standards that we are all very familiar with are road traffic regulations, which like health and safety laws are there for our protection, and are formulated to reduce the number of road traffic accidents and deaths whilst driving. The standards are clearly laid down and reasonably well understood, even though most of us have never read the actual statutory instruments. Competence is validated through a driving test, and ongoing compliance is audited by traffic policemen and an increasing array of high-tech cameras and gadgets.

Unfortunately, not everyone sees laws as helpful, and there is a natural human tendency to comply with the laws that we agree with and not to comply with those that we disagree with. The classic example must be roadside speed limits; can any of us honestly say we have never broken the speed limit? Our willingness to comply with a standard is often assessed by the likelihood of getting caught. We don't always comply with speed limit standards because having checked for the presence of police patrol cars, we assess that the chance of getting caught is very low, or at least that the benefits that arise from getting to our destination quicker outweigh the chance of being caught. In fact, our behaviour results from us weighing the benefits and consequences of our actions. If we perceive that the benefits significantly outweigh the disadvantages, then that is the action we are most likely to take, even if sometimes it means breaking the law. However, if the presence of the speed limit sign is re-enforced by the presence of speed cameras, our behaviour changes and we are more inclined to comply with the speed limit, at least until we think that we are past the camera, mainly because we think that there is a greater chance of being caught. In this situation, the speed limit represents a required driving standard, and the camera represents an audit or monitoring process that is intended to raise compliance levels (Figure 5.1).

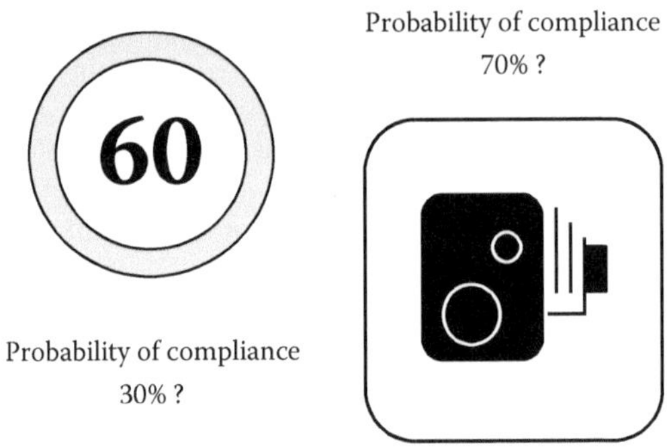

FIGURE 5.1 The car speeding analogy.

However, laws and regulations are not the only standards that we may be asked to audit. Some standards are issued by governments or their agencies (e.g. national or international standards organisations), which are not mandatory but which carry the power of law. 'Codes of practice' are an example of standards which are not laws or regulations but which can result in prosecutions if they are ignored. Taking the example of road traffic standards, the UK requirements are laid down in the Highway Code, but the Highway Code is not law and no one has been ever prosecuted directly for not complying with its requirements. However, many people are prosecuted every day for speeding or dangerous driving under the Road Traffic Act and the courts will cite a failure to follow the Highway Code as evidence of dangerous driving. Even though they are not law, codes of practice are standards of behaviour that can be used as the basis for an audit.

Although regulatory compliance is one common basis for audit standards, it is by no means the only one. Standards can also arise from non-governmental organisations such as industry associations. Similarly, engineering standards and specifications can be used as the basis for an audit. However, the most common application of audits in the SHE area is within organisations as a check against that organisation's own written procedures and instructions. The current tendency for organisations to carry out so-called safety audits where the auditor uses his or her own knowledge and experience as the baseline for compliance is actually a misnomer. This type of assessment should be described as an 'inspection' rather than an audit, because the reference standard is in the mind of the assessor and not a documented requirement.

It is possible for a handwritten checklist to be an acceptable audit standard. Many years ago, while carrying out my first major Level 3 audit on a large manufacturing site, one of the auditees complained to me that they were being assessed against something that they didn't know that they had to do. Of course, once this situation arises, other than being an awareness-raising event, this part of the audit becomes pointless because if auditees don't know what was expected, how can they be expected to comply? However, in compliance-level auditing (Level 1), if the auditor has consulted with the auditee in advance and has agreed the requirements in a written checklist, then this can in effect become the audit requirement for that occasion. The key point about any audit requirement is that both the auditor and the auditee must recognise and accept that this is the standard that is to be achieved.

What is clear, however, is that the definition of an audit talks about 'compliance with defined standards'. This concept is crucial, because in order to be defined, the standard or requirement must be written down somewhere, such that the auditor and the auditee can independently check the requirements of the standard, with the hope of having a common understanding of them.

Examples of some written requirements that might constitute the standard for auditing are

- Regulatory requirements
- Codes of practice
- Licences and authorisation documents
- Company policy
- Guidance notes
- Operating procedures
- Engineering specifications
- Pre-prepared audit checklists

The relevance of these requirements will vary depending on the type of audit being carried out. Operating procedures will be the most common type of standard when carrying out operational-level audits, whereas regulatory requirements and engineering specifications may be more commonly found to be the standards in Level 2 specialist audits. Occasionally, there may be a need to audit an activity where no readily available standard exists. As has been mentioned, in this situation it is necessary to establish an agreed standard or audit checklist that is acceptable to both auditee and auditor. Once the requirements are established and written down and agreed upon by both parties, then an audit standard exists and the audit may proceed and performance assessed against those requirements.

The great learning that arose from the use of ISO 9000 and the application of auditing outside the financial areas was that not all standards and procedures are auditable. In my own experience, although my company had plenty of binders full of instructions, on inspection we found that not many were in a form that could be audited. In order for standards and procedures to be auditable, they should clearly define the requirements and those requirements must be achievable. One apparent 'quality' procedure that I read said, *'To promote uniformity of working methods throughout, certain procedures fundamental to ________ Ltd., shall be implemented at all times without unauthorised deviation!'* Other than leaving the reader bemused and wondering what on earth he or she is meant to do, it begs the questions

- Which working methods?
- Which procedures are fundamental?

- Who is required to act upon this instruction?
- Who made this edict?
- Who can authorise a deviation?

Although we do not know what has to be done, why or by whom, the good news is that whatever it is that has to be done needs to be done all the time. At least the timescale part of the requirement is auditable, but with that exception, it would be impossible to audit such a statement. To be clear and unambiguous, the standards or their supporting procedures or instructions need to define

- What needs to be done
- How it needs to be done
- Where it needs to be done
- When it needs to be done
- Who is responsible for doing it

It may also be advisable to consider including

- Why it is necessary
- What happens next

An example of an auditable requirement would be as follows:

> Every accident resulting in occupational illness or injury, regardless of how trivial and all property damage accidents involving $500 or more loss will be investigated in accordance with company instruction SHE-3 by the line manager before the end of the working day or shift on which the incident happened.

It is essential that the scope of any audit clearly define what standards are to be audited. If these standards are new to the auditor, then it is essential that the auditor check the auditability of the standards before committing to carry out the audit.

6 Preparation

Although ISO 19011 provides little help in good auditing practice, it is very helpful in identifying what needs to be done in creating an audit programme and what preparation is required to carry out an audit. In my experience, the announcement of an audit is usually the signal for the start of some frenetic activity while the auditees try and get themselves into the best possible shape. It is human nature that most people like to be seen to do well in examinations, and audit assessments are no different. Some experienced auditors object to this sudden surge of activity before the audit in the belief that the outcome will be unrepresentative of the normal performance. The same auditors will accept the decision of an examining board that a graduate is qualified to a particular level but ignore the fact that the student's examination result was influenced by some perfectly normal last-minute revision or 'cramming'. It is my belief that the purpose of an audit is to stimulate improvement and if some of that improvement is self-motivated in advance, then the audit has started to achieve something even before the entry meeting. It is said that 'audits catch only the undedicated, bored or careless', so it seems to me that zealous preparation should be encouraged rather than discouraged.

However, the need for preparation does not reside solely with the auditee. The audit can be wrecked just as easily by an undedicated, bored or careless auditor in just the same way as by a bored or careless auditee. A successful audit needs advance preparation, irrespective of whether it is a 1-hour instruction compliance audit or a several-day major SHE management or specialist audit. The audit may be initiated either as a result of a specific one-off request or as a result of the operation of a rolling programme. In the latter case, it is quite possible that an approach from the auditor will be the first that local management know about the audit. It is quite common in these circumstances that the locals will not greet the announcement of an impending audit enthusiastically. It is part of the auditor's preparation to ensure that the exercise is seen in a positive rather than a negative light.

Usually, the auditor will have been approached either by the local senior manager or the manager responsible for running the audit programme. The audit programme manager will need to be satisfied that the audit is feasible, given the information that is expected to be available, the time and resources available to both plan and conduct the audit, and finally, the level of cooperation expected from the audited location. Assuming that the audit is feasible, then a qualified auditor will be appointed 'lead auditor'. The number of accredited lead auditors should be kept to a minimum. For the major SHE audits there is considerable benefit in keeping the lead auditor constant to minimise problems with the comparability of audit results. In the more common situation where the audit involves just a single auditor, then that individual also takes on the role of lead auditor. It is the lead auditor's responsibility to undertake the preparation for the audit and to select the other members of the audit team, if more than one auditor is required. This should always be done in consultation with the auditee, as some individual auditors may not be acceptable to the auditee. Any team needs to ensure that it has the right balance of experience, seniority, experience of the technology and regulatory requirements and, for international audits, a knowledge of the language. The last point is obvious, but many Anglo-American multinationals are surprisingly immune to this. I carried out an audit of one facility where the safety studies were all carried out in English, the equipment drawings were all in German and the operators were all Japanese. To aggravate the situation still further, all the controls in the Japanese plant were labelled in English. A real recipe for creating a safety incident!

The auditors will need to be flexible about their choice of audit dates, because auditing the finance department just when it is carrying out its month- or year-end closing accounts is unlikely to get a warm response. Furthermore, it will be essential to get the involvement of the senior management, and these people tend to be very busy and have diaries that are booked well in advance. As much prior notification as possible should be given, but this will vary for the type of audit. Notification for major SHE management system audits and specialist audits (Levels 3 and 2) is likely to be at least 6 weeks, whereas notification for a safety instruction compliance audit (Level 1) carried out by local staff may only need to be a few days. The notification should be in written form. For the major audits, this is likely to be a letter to the senior manager of the unit, whereas for a compliance audit, the use of a standard pro forma notification or even just an email is more common. The need for a written communication is to ensure that there is the maximum clarity about the subject and timing of the audit. The notification should also establish some means of feedback, usually verbal, to confirm that the message has been received and understood. Irrespective of the type of audit, the notification should contain the following information:

- Scope of audit
- Date of audit
- Names of auditors
- Outline of the audit programme
- Documentation requests
- Request for the name of the unit's appointed audit coordinator (i.e. the person who will coordinate the audit on behalf of the auditee)

(An example of an audit notification letter is incorporated into Appendix A1.4.)

The request for documentation in advance is actually a matter of personal preference. Many of the well-established commercial SHE auditing systems including ISO 19011 (Section 6.3 identifies the need to conduct documentary reviews prior to the on-site audit activities) make it an absolute requirement that large quantities of documentation should be provided to the auditor in advance of the audit. In my opinion, this is largely unnecessary in a major audit, as quite often these do not get read in advance and often these requests just become a chore for the auditee and another opportunity to devalue the real purpose of the audit. What is important is to notify the auditee of what documentation may be required during the audit and to see copies of the last audit report and policy statements in advance in order to give the auditor sufficient feel about what to expect.

Preparing the audit team, if it involves more than one auditor, is the responsibility of the lead auditor. It is his or her responsibility to ensure that the audit process is conducted effectively and that reports are written in a timely manner. The lead auditor must ensure that appropriate audit checklists or protocols are available and in the possession of the audit team prior to the audit. The lead auditor must ensure that he or she is properly prepared for the entry meeting and that the audit programme meets the auditor's as well as the auditee's needs. Decisions about the method and style of reporting, together with questions about whether there is any need to quantify the audit result in some measure of compliance, need to be resolved at this stage.

For audit teams of more than one auditor, I have found it extremely beneficial to prepare an auditor's manual that can be given to each auditor in advance and contains such things as

- Audit notification letter and other related communications with the auditee
- Audit scope
- Entry meeting presentational material or notes
- Audit programme
- Location layout plan (for large and complex offices or factories)
- Organisation chart of the management of the unit
- Previous audit reports
- Auditor's guidance notes or rules
- Checklists or protocols
- Quantitative reporting process, if required
- Blank copies of auditor's working papers
- Target numbers of discussions to be carried out
- Accommodation arrangements (if required)
- Details of any special rules or requirements of the local management, such as confidentiality agreements and special protective equipment requirements

(Also, see Appendix A1.6)

This information is normally in a photocopied format and contained within a suitable loose-leaf binder.

It is also normal for auditors to wear name tags at all times, identifying them by first name as SHE auditors. This helps employees, who may be unaware that the audit is going on, to recognise the auditors and address them by name. It is small details such as this that can help remove some of the threatening elements that so many people still associate with being audited.

The lead auditor must assume responsibility for the health and safety of his or her own team. He or she should check in advance whether there are any special health risks on the location to be audited and ensure that the remainder of the team is made aware. This is particularly important when auditors are visiting manufacturing plants, construction sites, farms or laboratories. The issue is not purely one of the health and safety of the auditors, but if there is a local requirement that everyone should wear hard hats and the auditors are seen not to comply with this, then the whole credibility of the audit will be undermined. It is critically important that auditors are seen to 'walk the talk' and set a good example.

Although most enlightened managers would agree that auditing plays an essential part in encouraging compliance with a set of legal or corporate requirements, that universally beneficial feeling soon evaporates once the long list of corrective actions starts to be unearthed. The biggest danger of auditing is that it becomes a critical process that generates a huge additional workload for already beleaguered managers. In undertaking preparation for the audit, the auditor should be sensitive to the reaction that his or her presence will have. There is an argument that on-the-spot, unannounced audits will result in a more realistic assessment of compliance. However, this approach will also generate the greatest antagonism and reinforce the feeling that the audit is attempting to 'catch them out'. On the other hand, there is an opposing view regarding the provision of ample warning of audits: that extensive prior notice gives the auditee time to 'fix' some of the problems and catch up on some overdue actions, hence creating an artificial view of the level of compliance. This may well be the case, but it must be remembered that the purpose of an audit is not just to identify the percentage of compliance but more importantly to identify opportunities for improvement. Surely, the initiation of some remedial activity before the arrival of the auditor means that some improvements have happened purely as a result of the advance warning of the audit, and this can only be good news. On balance, therefore, I recommend that auditees should have good notice of intentions to carry out an audit, as this helps maximise the positive benefits of the audit. That is not to say, of course, that there is no place for unannounced audits, but these should be used judiciously and as an exception rather than the rule.

Before conducting the audit, it is essential that everyone understand the scope of the audit. The arrangements for agreeing on the scope will vary depending whether the audit is Level 1, 2 or 3.

In the case of Level 1 compliance audits, where the auditor is trying to ensure that the actual aspect of SHE practice is carried out as specified in a particular standard such as a statutory regulation or company procedure, there should be an audit plan. The plan will specify which audit should happen and when. The value of planning ahead with audits is that it ensures that both the auditors and auditees understand the scheduling of when audits are due and allows for the effective utilisation of resources. The existence of an audit plan will also significantly increase the likelihood of audits being completed on time and avoid the rolling over of incomplete audits to some ill-defined date in the future. It can also ensure that times of peak workload are avoided.

Preparing an audit plan (Figure 6.1) needs careful thought if it is not to be purely a sequencing operation, as different audits will need to be done at different frequencies. Auditing activities which are relatively infrequent, such as ground and groundwater monitoring, may require auditing only every year or two, whereas those regulations and procedures that relate to safe systems of work may need to be done monthly or, on some hazardous installations, even more frequently. The decision regarding individual audit frequencies for compliance level will require management judgement, which takes into account the risk of the activity, the consequence of noncompliance and the previous audit history.

When audit frequencies have been agreed on, the plan can be put in place to ensure that the audit workload is spread evenly across the available time, taking account of such things as holidays and peak work activities, like stock checks, major maintenance

Standard	Jan	Feb	Mar	Apr	May	Jun	July	Lead Auditee	Lead Auditor
Regulatory Requirements									
Control of Substances Hazardous to Health Regulations		*						JMcC	SWP
Waste Management Regulations				*				PJL	RFG
Risk Assessments									
Company Requirements									
Procedure S-001 – Permits to Work	*		*		*		*	DCS	TIK
Procedure S-014 – Control of Modifications						*		WH	JLM
And so forth									

FIGURE 6.1 Example of an audit plan.

overhauls and production downtime. A useful format for audit plans is a rolling 1-year or 5-year matrix which is updated every 6 months or annually. The plan should identify the regulation or procedure to be audited, include the initials of the lead auditee and auditor, and usually indicate the month in which the audit is to be completed (Figure 6.1). This timing gives some flexibility to the auditor and auditees to agree to a suitable time within that month when they can all plan to be available.

The use of this rolling matrix form of audit plan means that both auditors and auditees know what standards are to be audited in the immediate future. Consequently, audits should not normally crop up as a surprise unless there has been some important learning event or gross noncompliance has been found in a similar area.

Defining the scope of compliance audits is not limited only to the sequence of which audit happens when. It is important to also recognise the breadth of the audit. For example, if the audit is to check compliance with the ergonomic standards within a retail organisation, is the auditor expected to audit across the whole organisation:

1. All the shops in the company?
2. All the shops in a particular town or locality?
3. All the shops using computers excluding computerised cash registers?
4. Distribution centres and offices as well as shops?
5. Only shops provided with mechanical handling equipment?

Likewise, in manufacturing concerns, the audit may be limited to certain departments or shifts.

There can be efficiency benefits from auditing a small part of a larger concern. These arise from the fact that not only does the audit take less time but also it is often the case that a nonconformance in one department may be repeated elsewhere in another similar department. Provided that the audited organisation shares and applies the audit findings throughout all departments, this can prove an effective use of time. However, the frequency of auditing will need to be increased to ensure that sharing occurs. So, for example, an organisation with five similar departments that might need to be audited for compliance with the Display Screen Equipment Regulations every 2 years might choose to audit one department only but to do a different department every year instead of doing the whole organisation every 2 years. In these circumstances, when the learning from an adjacent department is shared with the other four departments, the receiving manager can review his or her own practices and identify whether he or she has similar noncompliances. It is often easier, less confrontational and certainly less embarrassing for managers to recognise and act on their own department's shortcoming, rather than have some stranger come and tell them that they are out of compliance.

The preparation for Level 3 management audits is a whole different matter. Although it is necessary to have an audit plan to define when the management audit will happen, the plan itself will not help in defining the scope of the audit. It must be remembered that the SHE management audit is a broad scan of whether the management team has all the appropriate systems in place to ensure that it operates with an acceptable level of SHE loss prevention. If the organisation has a good track record in SHE management, then it is quite likely that it will know all the things that require doing. However, because the purpose of the SHE management audit is to ensure that all the necessary areas are adequately covered, then a location being audited for the first time may not be aware of all developments in regulation, industry or company requirements. In these circumstances, the location's management team may not recognise the full audit scope in advance, because it is unaware of the requirements. In the words of Professor Trevor Kletz, 'They don't know what they don't know.' In these circumstances, the detailed scope of the audit may not be known in advance or when the audit is scheduled in the plan. This type of rather ill-defined scope is common when smaller companies invite external consultant auditors into their operations and usually reflect committed and concerned managers who are less interested in knowing their level of compliance but are much more interested to learn what they have to do to comply with the law. With the exception of the aforementioned example, in most other cases, the scope of the audit should be agreed well in advance (say, 1 month before the audit) as people's time needs to be booked and auditors with the appropriate skill found. A list of SHE aspects that may need to be considered for inclusion within the scope is found in Appendix A1.1. In the event of disagreement, the scope should err on the side of including aspects where doubt exists about their relevance, as the audit itself will be the final arbiter and will identify whether the aspect is relevant to that location or not. Experience has shown that subjects that tend to be overlooked in the auditee's version of the scope include such things as the safety of employees' travel, product safety, the safe loading and unloading of goods, the safety of activities carried out by agency employees or contractors and so forth. A classic example of how perceptions of applicability of different aspects of safety may vary occurred during an audit that the author carried out in North Carolina. The pre-audit scope had suggested that one of the aspects to be considered was the safety of railways. The audit manager at the site suggested that this aspect of safety was not relevant to their operation and so that aspect was removed from the scope. Imagine the auditors' surprise when, on arrival at the company for the first time, the security guard directed them to the main offices with the words 'Follow the railway lines as far as you can go.' It transpired that the audit manager really meant that they had no locomotives of their own, but rail traffic was still a potential hazard on the facility.

The scope for specialist (Level 2) audits will also need to be agreed on in advance, in a similar way to that described for management audits.

The fact that a particular aspect was or was not incorporated into a previous audit is no guarantee of its relevance to the scope on this occasion, as circumstances and standards change with time. In particular, new equipment may be purchased and old equipment scrapped, leading to changes in the environmental, health or safety requirements. So the auditor must always review an old scope with the location management to ensure that it is still relevant.

It is quite common to use proprietary audit systems for management audits, such as the International Safety Rating Scheme (ISRS) from Det Norske Veritas. Other proprietary audit systems are available for specialist and regulatory compliance audits. Systems such as ISRS have been successfully applied across a wide range of industries internationally, but because of their broad nature, the auditee requesting the audit must ensure that not only is the scope of the proprietary audit fully relevant to their situation but should also ask if any aspect has been missed. This is particularly important in industries and organisations where there may be special needs, such as the food manufacturing industry or hazardous nuclear or chemical installations. Some proprietary audit systems operate only through the use of well-trained, experienced and accredited auditors, but many of the less expensive computer-based audit systems may be used by inexperienced managers who assume that the software will cover the full scope that they require. It is always necessary to check with any proprietary system what part of the scope provided is relevant to your needs and what is irrelevant or missing.

Once the scope is agreed, many auditors insist on being provided with a very comprehensive list of paperwork for pre-reading before the audit commences. This is commendable in theory, but the author's experience is that much of this material fails to get read in advance, and the demand for this pre-reading material purely creates an unnecessary clerical workload for the auditee. Demands for advance copies of documents should be kept to the absolute minimum required to ensure that the auditor is sufficiently knowledgeable about what needs to be done on the site to comply with the audit topic. It is sometimes useful for the auditor to provide a brief list of materials and systems that he or she will require to see during the audit but not to request multiple copies in advance.

The consideration that the auditor should give to the auditee cannot be overemphasised. The fewer demands that the auditor makes in advance, the less 'policeman-like' the audit will appear and the more likely it will be that the auditees will learn from what they perceive to be a positive rather than a negative experience.

One of the requirements of the new international standard on occupational health and safety management (OH&S) (ISO 45001; see Chapter 32) is that 'the organisation shall establish a process to ensure effective participation and consultation in the OH&S management system by workers at all levels and functions of the organisation'. It should be noted that this requirement also relates to worker consultation and involvement in the audit process.

In summary, the preparation for the audit should include

1. Statement of what is to be audited
2. Audit scope, including the parts of the organisation to be audited
3. Dates and locations for on-site audit
4. Time and duration of the audit
5. Selection of appropriate audit team
6. Presence of any trainees
7. Identification of auditee's audit manager/coordinator
8. Agreement on working and reporting languages
9. Audit report style
10. Logistics (meeting room booking, travel, hotel rooms, etc.)
11. Obtaining relevant previous audit reports and action lists

(Also see Appendices A1.1, A1.2, A1.3 and A1.4.)

7 Protocols and Checklists

We have already referred to an audit as a 'systematic examination'. In order for the examination to be systematic, it requires clear standards and instructions followed by a defined and prescribed means of evaluating the compliance and adequacy of those procedures. To achieve the latter point requires the use of checklists or protocols. As mentioned earlier, airline pilots ensure compliance with their wide range of pre-flight checks through the use of a range of checklists. Using this method, the cockpit crew minimise the chance of overlooking some important detail. Likewise, an audit without defined requirements or a structured checking process is no more than an inspection. Inspections have some value, but their quality is totally dependent on the experience, knowledge and thoroughness of the person carrying out the inspection. It is impossible to say whether the difference between two inspections carried out by different individuals represents a real significant change in performance or whether it is just that the inspection quality differed. Many of the workplace inspections that are carried out today are incorrectly referred to as audits. That is not to denigrate workplace inspections, as these processes are hugely beneficial and focus on the most important area of SHE control, which is human behaviour; but they are not true audits. The big difference between audits and inspections, therefore, is that the inspection is a limited examination through observation, and the audit is a thorough examination against a defined standard or requirement.

It does not matter how broad or narrow the scope of the audit is, effective preparation is essential and cannot be avoided.

As with any investigative process, the success of any SHE audit lies fundamentally in asking the right questions. The police detective will never crack his case if he does not ask the right questions of both himself and the accused. This skill is also paramount for the effective auditor. The trick to asking the right questions lies in the quality of the preparation. When auditing a procedure, I have seen many examples of the Level 1 auditor attempting to scan through the procedure for the first time while trying to formulate suitable questions and at the same time listen to the auditee's response to the previous question. As any good home decorator will tell you, success is all about the quality of your preparation. It will be immediately obvious to the auditee if the auditor is ill-prepared. This will not only personally compromise the auditor but will undermine the validity of the whole audit system.

No audit should be attempted without a checklist of some sort; otherwise, it is unlikely that the auditor will remember to examine the full breadth of requirements. There are many sources of checklists and protocols ranging in length from a few lines to 300 or more pages. (An example of a protocol is available in Appendix A2.) Many of these proprietary systems are available commercially. Some commercially available products focus on the environment or safety or one aspect such as fire management or crisis management. The ideal situation for organisations is to develop their own checklists or protocols which address compliance to their own particular standards and instructions. However, the warning is that there should not be multiple different versions of the checklist for any one standard within the same organisation, as the relative consistency is important.

Typically, there are two general types of audit preparation: these are 'bespoke' preparations and 'pre-prepared' preparations. Each of the two approaches has its own benefits and disadvantages. The bespoke preparation done by the individual auditor has the benefits that the auditor checks his or her own understanding of the requirements and produces his or her own checklist which this auditor understands. However, this auditor's interpretation may differ from that of another auditor preparing a checklist for the same procedure but at a different time. This has the disadvantage that the production of individualised checklists can result in varying audit standards when different auditors working to different checklists carry out subsequent audits.

Pre-prepared checklists can come in many different formats. The simplest form is using a checklist prepared by a previous auditor. It is quite common to use proprietary audit systems, such as the International Safety Rating Scheme (ISRS) from Det Norske Veritas for management audits. Other proprietary audit systems are available for specialist and regulatory compliance audits. Systems such as ISRS have been successfully applied across a wide range of industries internationally, but because of their broad nature, the auditee requesting the audit must not only ensure that the scope of the proprietary audit is fully relevant to their situation but also ask if any aspect has been missed. Unfortunately, some pre-prepared checklists are so prescriptive that sometimes the auditor does not understand what is behind the question, or alternatively the auditee knows in advance precisely what questions will be asked and how to get the optimum result.

Generally, protocols are well liked by auditors and auditees alike, ensuring a robust and effective audit, but they can be abused. During one audit using a commercially available protocol, an auditee challenged the auditor: 'Why am I being asked this question? I didn't even know that it was a requirement!' These situations are rare but very damaging if the auditee is correct. The audit must cover only the known standards and requirements that relate to that particular enterprise and should not inadvertently slip into areas of the auditor's experience or best practice which do not relate to that audit situation.

The use of either bespoke or pre-prepared checklists is very much a matter for selection by the management team or the auditor, but it should be done on the basis of needs rather than some whim. Figure 7.1a and b summarises the advantages and disadvantages of the two systems.

Whether auditors use a pre-prepared protocol or whether they prepare their own bespoke checklist does not affect the method used to develop the questions, as the same philosophy is used in both cases. In either case, the auditor is trying to establish the answers to a set of four generic questions. In the words of Rudyard Kipling, 'I know four wise men, they are who, what, where and when.' The auditor must explore a variation of Kipling's theme: the auditor needs to understand *who* is required to do *what*, *when* and *how*.

Benefits	Potential disadvantages
Specific	Dependent on auditor's interpretation of key elements
Good auditor understanding	Time-consuming for auditor to prepare
Checklist is usually brief	Different auditors may assess compliance differently

(a)

FIGURE 7.1a Bespoke checklists (protocols).

Benefits	Potential disadvantages
Standardisation of audit quality from one audit to another	Protocols can be very long
Eliminates some auditor variability	Auditor may not fully understand the protocol's questions
Auditee knows in advance what has to be done	May be unspecific in relation to particular procedures or regulations
Economical on audit preparation time	Possible for the auditee to 'brush up' on the questions in advance
Protocol usually prepared by an 'expert'	Not always obvious if the protocol exceeds or does not cover all of the audit scope
Can be an aid to verification	
Comparative data often available for benchmarking audit results	

(b)

FIGURE 7.1b Pre-prepared checklists (protocols).

WHO

'Who' identifies accountability. This will usually be the person that the auditor needs to talk to in the first instance. Inevitably, the audit trail will lead on from this individual to others who may be carrying out delegated tasks. The person(s) that the procedure is directed at will make an excellent starting place for your enquiries.

WHAT

'What' is the main subject matter of the audit. In Level 3 (management) audits, this may be a broad general statement such as *'Ensure that procedures are in place to prevent exposure to chemical XYZ.'* In the Level 1 (compliance) audit on the same subject, the procedure may require that *'cartridge-type respirators, chemical suits, PVC gloves and face hoods are to be worn when handling chemical XYZ'.*

Although both these procedures are intended to address and control the potential hazards associated with chemical XYZ, in the first case (the Level 3 management procedure), the audit check for what has to be done might be as follows:

Ensure that an up-to-date procedure exists to control the hazards of handling chemical XYZ and ensure that relevant employees are suitably trained and so forth.

In the compliance audit, the check for what has to be done might be as follows:

Ensure that cartridge-type respirators, chemical suits, PVC gloves and face hoods are available and that the wearers are trained in their use.
Ensure that these protective equipment requirements are used on all occasions when handling chemical XYZ.

It should be clear that although these two procedures relate to the same topic, because they are aimed at different levels in the organisation, the auditor's checklist for 'what has to be done' will be different. In the first case, the auditor will be looking for evidence that the system exists, which will entail discussions primarily with management, whereas in the second case, the auditor will be looking for evidence that workers and other people affected are complying with the protective equipment requirements of those systems. In the latter case, the auditor's focus will be on whether employees actually wear the prescribed personal protective equipment when handling chemical XYZ. The preparation of the checklist relating to 'what has to be done' will therefore be different for the management (Level 3) audits and for the compliance (Level 1) audits, even if they nominally cover the same subject.

HOW

The preparation of checklists to cover 'how' the task is to be done will also vary depending on the circumstances. In many cases, modern procedures will specify criteria for what has to be done and in what sequence, but they may not specify precisely how it is to be achieved. For example, a laboratory procedure may require a trained technician to carry out a titration of a certain substance. It may be reasonably assumed that a trained and competent laboratory technician does not need to be told step by step how to carry out a titration, as this will have been part of the technician's basic training and is now assumed to be a skill. Similarly, electricians trying to diagnose a fault can be given general principles of safety and guidelines but cannot expect to have a pre-prepared procedure on how to diagnose every conceivable fault that they may come across in their working lifetime; they will need to use their skill and judgement on how best to diagnose the problem.

We have all experienced the situation of purchasing a new piece of equipment for the home, where after hours of frustrated failure we finally resort to the old adage 'If all else fails, read the instructions!' This may be a common, albeit questionable, approach to the erection of self-assembly furniture, but it is most certainly not the approach if you are dealing with more hazardous electrical or garden machinery, where the consequences of your misunderstanding may be considerably more severe. In preparing the check questions for 'how the task is to be done', the auditor must appreciate the consequences of failing to meet the requirements and suitably adjust the checklist to reflect which of the requirements are most important in terms of injury or environmental protection. The auditor will in effect be carrying out a simple risk assessment.

Although many procedures may be unspecific about exactly how the task is to be done, when detailed step-by-step instructions are provided, the auditor should initially assume that following this stepwise process is important. In these circumstances, verifying the events laid down in the procedure may form a part of the auditor's checklist.

WHEN

The fourth step in preparing the checklist is to address 'when things should happen'. This may relate to when in terms of a sequence of events that things need to be done in order to complete a task, or it may relate to when the task has to be done on a calendar basis. This is very common for equipment examinations that occur on a periodic basis. Quite often these tasks are specified by law and may cover such things as vehicle safety checks, boiler and air receiver inspections and so forth.

Frequently, when a sequence of events has to be followed, it is common for operators to use a 'tick off' checklist, and in these circumstances the auditor may have to verify only that the checklists have been completed and that has been done in the right order. However, requirements to ensure that certain events happen at the right frequency, such as the inspection and maintenance of portable electrical equipment or fire extinguishers, will require the auditor to examine historic records and inspection schedules to ensure that the specified work was carried out at the appropriate time.

Having completed the initial audit checklist by asking the question 'Who should do what, when and how?', the auditor must remember that auditing is not something that is performed by a robot. The auditor must now check that compliance with the newly prepared audit checklist will ensure that the intent of the procedure has been fully met. This often requires the auditor to exercise some judgement, and although this is essential, it is in the application of an individual's judgement that there is the potential for variability in the final audit results because different auditors may apply their judgement in different ways depending on their own skills, experience and bias.

I was carrying out a compliance audit into a procedure relating to management communications. The procedure identified that the senior manager was responsible for communicating important safety and business information to his immediate subordinates every Monday afternoon. Records had to be kept of who attended the communications meeting and what was discussed, and these records had to be kept in a specific file. The audit identified that all aspects of the procedure were being complied with and recordkeeping was immaculate. I then asked myself, 'What was the underlying purpose of this communications procedure?' Was it to have effective communication throughout the organisation to those people who need to know, or was it to have an exemplary system of records of communication? I concluded that it was the former and the senior manager concurred. I then adapted my audit checklist to include a question that was not derived directly from the procedure. The question was, 'Do all employees receive and understand the communication that is relevant to them?' Armed with this new question, I rapidly found that some employees did not receive any communication; others thought that they had received some, but didn't think it important to them; and the third and largest group had been told things that they didn't understand. Clearly, although the prescribed system was working according to the procedure, this had not resulted in effective communication. So it is always necessary to ensure that the answers to the auditing checklist meet the intent of the procedure being audited.

An easy way of preparing an audit checklist is first to obtain the most up-to-date copy of the procedure or instruction to be audited. Then, using a coloured highlighter pen, go through the document and highlight the phrases relating to who should do what, when and how. This process is not only easy and quick, but it avoids the other *bête noir* of the auditor, which is getting caught out by not having read the most recent copy of the procedure before commencing the audit. Just using a highlighted copy of the procedure as the basis for your audit is not effective, nor is it likely to convey an atmosphere of professionalism. The highlighted phrases should be transposed into audit questions in a concise purpose-designed checklist. In carrying out this transposition, the auditor must ensure that the questions generated are posed in an auditable manner. Unless this is thought about in advance, the auditor may glean a great deal of interesting information, but he or she may not be able to conclude whether or not the auditee has complied with the procedure. The auditee may have argued a very plausible case for how the task is carried out, but that may bear little resemblance to what he or she is really meant to do.

The phrasing of any audit question should result in a response with a clearly auditable outcome. In the ideal scenario, the answer in the auditor's mind to the audit question should be an unequivocal 'yes' or 'no' relating to whether the auditee complies with the audit question. For example, an unauditable instruction might be as follows:

Factory effluent should be analysed.

First, this is a statement rather than a question. Auditing this requirement is likely to be inconclusive when trying to decide whether the factory is compliant in analysing its effluent because it raises a number of unanswered questions regarding the requirements:

1. Is the fact that a sample was analysed 12 months ago good enough?
2. What is the analysis looking for?
3. Who has to ensure this happens?
4. What is an unsatisfactory analysis result and what are the consequences?

The result of asking questions about this requirement will undoubtedly provide a lot of information, but it may not be conclusive. On the other hand, try posing the audit protocol question as follows:

Are effluent samples taken and analysed (daily/weekly/monthly) and do the analysis results always confirm that the factory is operating within its discharge consents?

The answer to this question is either 'yes' or 'no'. There is a clarity that is beneficial to both the auditor and auditee. Equally, if most of the time the factory is compliant but it has had one or two deviations, the auditor can conclude, 'The factory is mainly in compliance, with two out of twenty recorded noncompliances in the effluent discharge over the last 20 months.'

To identify where noncompliances exist, it would be even better if the question were further subdivided, because this question is checking compliance with both

1. The effluent sampling system
2. Factory compliance with its waste water discharge consents

It is possible that the sampling system is working well but that the factory is out of compliance on the effluent contaminants. So, a more specific audit process would ask two questions in the audit checklist:

1. Are audit samples taken and analysed (daily/weekly)?
2. Do analysis results show that the factory is operating within its discharge consents?

Factual and unemotional statements of this sort are essential when it comes to compiling the audit report and substantially reduce the chances of the auditor misinterpreting the situation. The avoidance of misunderstandings, observational errors and erroneous conclusions is essential in ensuring the credibility of the outcomes of the audit and greatly increases the chances that the audit recommendations will be adopted. If conflict or disharmony arises between the auditor and the auditees, then the value of the audit is almost immediately undermined and the likelihood that corrective actions will be adopted is reduced. The auditor must remember from the outset that it is the quality and incisiveness of his or her questions that will determine the effectiveness and acceptability of the audit. Time spent ensuring a relevant and thorough checklist or protocol will repay itself handsomely by the end of the audit.

For organisations that may have several hundred SHE procedures to audit, the prospect of preparing a large number of audit checklists can seem daunting. In most cases, audits will be conducted on a periodic basis, according to some pre-prepared plan. This means that the same audit will be repeated at some frequency that may be weeks, months or even years. In this case, it is advisable to retain the audit checklists in a file or database to minimise the time needed for preparing audit checklists on future occasions. The retention and reuse of these checklists will help to overcome the major shortcoming of bespoke checklists— namely, the variability in audit standard and outcome. If subsequent auditors are working to the same or very similar checklists, then the compatibility between periodic audits on the same procedure will be greatly enhanced.

Nevertheless, we must exercise some caution here. If the initial audit checklist is of poor quality or incomplete, reuse of this flawed list will perpetuate a poor standard of auditing. Second, the auditor must recognise whether the original procedure, against which the audit checklist was prepared, has now changed. Procedures are reviewed and evolve with time, and this means that the key requirements may also change. Consequently, even if an audit checklist already exists for the procedure to be audited, the auditor has a duty to ensure that it is still appropriate. If substantial changes are to be made to the existing checklists, then the auditor should inform the appropriate auditee, so that the auditee understands in advance that the audit results may differ significantly from those of the previous audits.

Checklists should normally be prepared in a format that enables the response to the audit question to be written alongside or beneath the relevant question, as shown in Figure 7.2. Having the audit checklist on a piece of paper different from that on which the responses are compiled frequently leads to missing some key points.

Auditor checklist			
Ref.	**Audit question**	**Compliance** Yes/No	**Comments**

FIGURE 7.2 Auditor's checklist.

Although it has been recommended that the auditor work primarily from the checklist, it is essential that he or she have a copy of the procedure readily available during the audit. If challenged during the audit, the auditor must be able to identify where in the procedure the questions that he or she is asking are specified. To aid this, it is helpful to cross-reference the audit checklist question with the paragraph or page number of the procedure being audited. This is the purpose of the column headed 'Ref.' in Figure 7.2.

In summary, the process for preparation of audit checklist is laid out in the Checklist Preparation Flowchart shown in Figure 7.3.

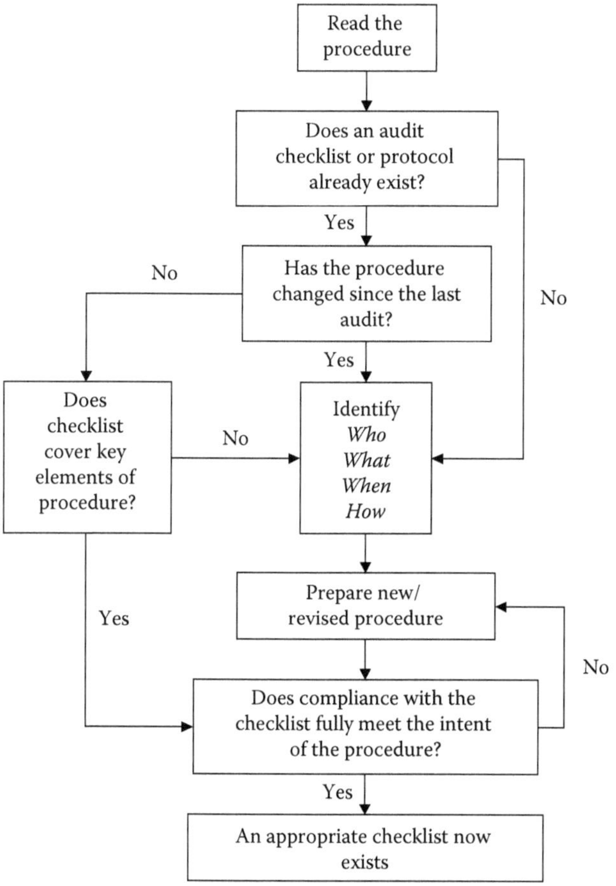

FIGURE 7.3 Checklist preparation flowchart.

In developing a set of generic questions that could apply to auditing the management systems, the auditor should consider the following:

1. Why does the standard or instruction exist?
2. What legislation applies and is it understood?

3. Are written procedures/instructions in place to ensure that the requirement is met?
4. When were the procedures last reviewed and updated?
5. Are all the responsibilities for managing and carrying out the procedures assigned? (Are the named people still alive?)
6. Are the people who are required to act on this requirement trained and validated?
7. Does local monitoring or auditing regularly assess the degree of compliance with this requirement?
8. What are the last set of corrective actions and what is their state of implementation?
9. What are the consequences of failure to comply with this requirement?

Armed with these generic questions, the experienced SHE auditor should be able to develop detailed checklists for most of the requirements of a management or specialist-style audit, irrespective of whether the system being audited is related to occupational health medical surveillance, control of ground contamination or arrangements for entries into confined spaces. The auditing principles are the same. Compliance auditing of detailed SHE instructions should simply look at the who, what, when and where requirements laid down in the instruction and probe to check whether the requirements specified in the instruction are being carried out.

However, the process for all audits can be streamlined and made more robust by always using the same style of questionnaire. This is particularly valuable if there are a large number of audits to be carried out against the same standards. In this case, there is value in developing a detailed list of questions and points for inspection and checking, so that all the audits are carried out to the same depth and level of detail, irrespective of which particular auditors are involved. Such a questionnaire is usually referred to as an 'audit protocol'. The best example of such a detailed protocol applied to a large number of audits is the ISRS system from Det Norske Veritas, which has become something of an industry standard in this area. The drawback of protocols, however, is that they can become an unthinking process that results in the auditor becoming a slave to the questions, rather than using them as a set of useful guidelines. My own approach is to invite the auditee to tell the auditors about 'how a particular system works' and I usually find that the response answers 60%–70% of the protocol questions. It is then necessary only to test the remaining questions from the protocol that have not been answered. The great advantage of the protocol is when it comes to recording responses, because these can easily be related to specific questions, minimising the auditor's note-taking but keeping things crystal clear when it comes to transposing the notes into a final report. When producing the protocols for right-handed auditors, it is advisable to print details on the left-hand pages only, leaving the right-hand pages free for note-taking. Usually, the annotated protocols from each member of the audit team are all retained by the audit team leader as part of the archive of working papers from the audit. These working papers can then be used not only for report writing but also for responding to queries later on, when memories of the reasons for certain recommendations start to become hazy. Working papers should normally be retained until after the next audit. The only real drawback of the protocol is that it does take some time to develop in the first place, but it is usually very efficient in the long term and does ensure that the audit assesses the organisation's actual requirements rather than some generic industry best practice that may or may not be totally relevant.

The use of pre-prepared protocols for assessing compliance with local safety instructions is not usual. In this case, the auditor carrying out the first audit should prepare a checklist using the guidance detailed above and then ensure that the checklist is archived for future use.

Many modern audit protocols and checklists come as computer files. This is fine when a manager is reviewing his or her own organisation's EHS performance from his own office, but sitting behind a computer screen while trying to discuss aspects of an audit with an auditee can be quite intimidating, and it is very difficult to take a laptop with you when carrying out a site tour. I always recommend that using handwritten notes on pre-printed computer printouts conveys a much more open approach and is less threatening for the auditee.

8 The Entry Meeting

The entry meeting is usually the first step of the actual audit. If the audit includes more than one auditor, especially if the auditors have not worked together previously or if one of them is a trainee, then the entry meeting will have been preceded by an auditors' meeting to ensure that the members of the audit team are all prepared and understand the process that is to be used.

The purpose of the entry meeting as defined by ISO 10011 is to

- Introduce the members of the audit team to the auditee's senior management
- Review the scope and the objectives of the audit
- Provide a short summary of the methods, procedures and programme to be used to conduct the audit
- Establish official communication links between the audit team and the auditee
- Confirm that the resources and facilities needed by the audit team are available
- Confirm the time and date for the closing meeting
- Clarify any unclear details on either side

The meeting will be chaired by the lead auditor and should be attended by the local senior management and whoever else the management team requires. It will be an immediate test of commitment to see who attends the meeting and who sends excuses or deputies. The meeting can follow a very similar pattern for all audits. In the interests of efficiency, the lead auditor may well have a set of standard presentational materials for this purpose as this is often the first opportunity that the auditors have to establish a professional image. It is important that the meeting is short and businesslike and should not exceed 15 minutes. The lead auditor must emphasise, from this very first contact, the positive nature of the audit and that the purpose is to help the process of continuous improvement rather than to be critical.

The presentational material for the meeting might include slides on the following topics:

- Purpose of the audit
- Names and background of auditors
- Audit scope
- Audit programme
- Logistics and arrangements
- Reporting arrangements and exit meeting

It is essential in the case of audits taking several days that the entry meeting confirm the ongoing communication process that will exist during the on-site phase of the audit to ensure that the auditees remain aware of the general finding of the audit team. It is always useful, if there is either very good or very bad news to communicate, that this does not suddenly appear at the exit meeting. The Plaudit 2 process using Post-it® Notes displays, described in Chapters 17 and 28, is one effective way of doing this. Alternatively, a lunch discussion or short end-of-day meeting between the lead auditor and the auditee's representative can achieve the same level of communication. During Level 3 management or process safety audits, be cautious about committing to daily long feedback meetings, as these can easily absorb a very significant proportion of the available auditing time.

9 Area Familiarisation

In a major SHE management or specialist audit, which typically will go on for 2 or more days, it will be necessary for the auditors to get a feel for what type of work goes on at the facility. This is especially important where the facility is a manufacturing, construction, farming or laboratory location, because this will be the first indicator to the auditor of the degree to which the standards are actually being implemented. The area familiarisation tour is not just an interesting sightseeing jaunt but is a key step in the audit process.

The tour will give the auditors an initial first impression of not only what the organisation does but more importantly how management and workers apply general standards, such as housekeeping, hygiene and road traffic requirements. These early observations can give an insight into how seriously people who work at the facility respond to the accepted norms and where bad practices might be creeping in. Information gleaned at this stage may well change the auditors' views and cause them to refocus the questions that will be posed during later discussions. The auditor should always take a copy of the site plan on the tour so that he can annotate it with reminders of places that he feels need to be revisited. It is also worth taking careful note of the route taken during the tour and, in particular, registering those areas that your guide studiously avoids; those may well be just the places that you want to come back and revisit. An essential piece of equipment at this stage is the auditor's notebook or clipboard, as your guide may not be one of the people closely involved in the audit and therefore may be somewhat more forthcoming in his or her comments than the boss would be. However, avoid the temptation to make the tour too detailed at this stage; there will be further opportunities for more informed plant inspections later in the audit. Taking a camera during the familiarisation tour can be very helpful, but always check in advance that your host will permit you to take photographs. This is especially important where the area could have commercially sensitive activities and equipment or potentially flammable atmospheres present. Remember to 'walk the talk' whenever touring the area and ensure that you, the audit team and your host all observe the safety and health signs and requirements while walking around. As a courtesy and for your own safety, if you want to explore some parts of the facility that are outside the designated tour route, always check with your host/tour guide before doing so, as you may be inadvertently exposing yourself to some hazard that you are not aware of.

10 Audit Observation Skills

To acquire knowledge, one must study; but to acquire wisdom, one must observe.

Marilyn vos Savant

There are a number of ways that the auditor can obtain information. These include discussions; reviews of documents, databases and drawings; and also information gained from observations of activities and conditions. During the familiarisation and later tours of the facility, auditors will be using all their senses of hearing, smell, touch and sight. One of the most important elements of auditing is the skill of assimilating information through observation. But we have spent all our lives looking at things; surely there is little that we have left to learn about this skill that we have spent a lifetime perfecting? We may not always believe everything we hear or read, but like the disciple Doubting Thomas, who would only believe that Christ had risen from the dead by seeing it for himself, when we see something first-hand for ourselves, we tend to believe it. Unfortunately the old adage 'Seeing is believing' does not always hold true for the auditor. Imprecise witness observations can be a common problem of identification evidence in our courts of law. The quality of a witness's visual evidence can depend on a range of factors. Take, for example, a person witnessing a handbag snatch and who subsequently reports seeing a man snatching a lady's red handbag. How reliable is that observation? If the witness was a woman, with good knowledge of handbag fashions, she might have reported it as a shoulder bag, whereas a male observer might just refer to the bag as a shopping bag or small case. Was the bag really red, or could the witness suffer from red–green colour blindness? Was the apparent snatching an intentional crime, an accident or acting as a part of a film? The skilled legal advocate will often use a myriad of arguments to suggest why apparent witness reports may not represent the truth or at least be mistaken perceptions. The thing to remember in this legal analogy is that the legal advocate has a clear objective relating to witness observations. If it helps the advocate's case, the advocate will want to argue that the observation represents the 'truth', whereas if it hinders his or her case, the advocate will want to find reasons the witness is mistaken. These apparently opposing views of what represents the 'truth' in what has been seen remind us that what we see is actually what we perceive. When watching a magician, we may see eggs disappearing into thin air or white doves appearing from inside a folded handkerchief, but as educated adults we know that eggs and doves cannot just vanish or appear. It is an illusion that the magician intends to create, and our eyes and minds have been deceived; the magician has persuaded us to see what he or she wanted us to see. Very often we see what we want to see rather than reality and that is certainly true in the case of audit observations.

FOCUSED LOOKING

Overfamiliarity is often cited as a reason apparently obvious things are overlooked. This is often the case in accident investigations, where with the benefit of hindsight, the investigator identifies what went wrong and how it could have been avoided. But the investigation has to focus on only the narrow range of health, safety or environmental issues, whereas the victim of the accident was thinking about the urgency of the job, the next task, an argument with the boss, the children's education, who would win the match on Saturday and 101 other things. 'He couldn't see the wood for the trees.' To prevent the injury, the victim needed to focus his or her mind on the things that were important at that particular time. Likewise, the auditor needs to plan his observations to ensure that he or she looks for what is important in relation to the topics that are being audited. It is the job of the auditor to be able to see the wood in the trees. Like any other successful activity in life, we need to be prepared, and audit observations are no different. It is very common to see what you want or expect to see. In training auditors to observe, I often use clips from comedy films. The trainees enjoy the humour but are then asked questions about details of the film set or what the characters were wearing. It is unusual for them to answer many questions right. The trainees are then shown a second film clip but are told what to look for in advance. Perhaps not surprisingly, they tend to get all of these questions right. The moral of this exercise is that it is easier to get the right results if we plan ahead and know what we are looking for. This is another area where audits differ from inspections. The audit is a 'focused look' at some particular aspect of health and safety, whereas inspections tend to be general 'unfocused' observations.

Focused observations sound easy, but the auditor must be aware of the pitfalls. The auditor must do sufficient advance work to know where he or she may need to go to look for a particular example of audit compliance. For example, there is little point in looking for evidence of compliance with ionising radiation standards if the organisation does not use any radioactive sources.

Even when we know what we are looking for, it is not always easy to understand what we are seeing. Our eyes may be deceived by what we know as an optical illusion, or it could be that our eyes can assimilate the information but our brain is not programmed to interpret it properly. For example, look at the symbols shown in Figure 10.1. Our eyes can clearly see that there are seven symbols and that they are Chinese characters, but the brain cannot properly interpret the characters unless we are trained to interpret Mandarin Chinese.

让 我 们 认 识 系 统

FIGURE 10.1 An instruction in Mandarin.

In fact, if our brain was correctly programmed and we could read Mandarin, we would understand that these symbols say, 'Tell us about the system.' Actually, that is not quite true; there is no word in Mandarin for 'system', so what it actually says is, 'Tell us about the filing', which of course is not the same thing at all. This reminds us that not only do we have to be able to interpret what we are seeing but we must recognise that how we interpret what we are seeing may be a little different from what others may see.

Even when we understand what we see, we will still be influenced by the circumstances or surroundings in which we observe things. The context is crucial to understanding what we are looking at. Study this sequence of stencilled letters; the character in the square is seen to be the letter 'B'.

Now look at the sequence of numbers; the number in the square is seen to be the number '13'.

If we now display the two groups of characters together, it can be seen that the character in the square is the same, and how it is viewed depends on whether it is seen in the context of a line of numbers or a line of letters.

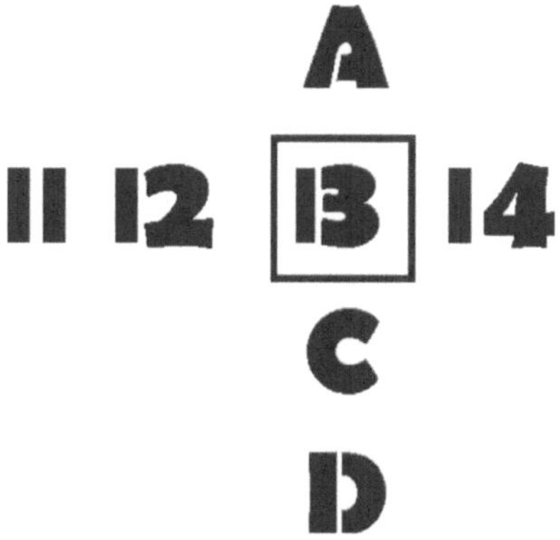

This simple example demonstrates how what we see is affected by the context in which we see it and is the principle behind many successful advertising campaigns. In the auditing context, this means that we can often observe something as significant if it is outside the context that we normally expect to see it, but the same fault observed in its usual surroundings may go unnoticed.

Reading is also subject to optical illusions. Try speaking aloud the *colours* of the following set of words as quickly as you can.

RED YELLOW PINK GREEN

BLUE ORANGE BLACK RED

PURPLE YELLOW GREEN

It is actually quite difficult, because the right side of your brain tries to recognise and speak the ink colour, whereas the left side of your brain insists on still trying to read the more familiar words. These simple examples show that, contrary to most people's expectations, seeing things for yourself is no guarantee that our brains are not playing tricks on us.

An abnormal context leads us to another situation where the auditor's observations can be important. Look at the photograph in Figure 10.2 and try and determine what the auditor might observe.

FIGURE 10.2 Auditor's observations: a green field.

Inexperienced auditors tend to focus on the state of the fencing in the foreground. The experienced auditor tries to understand the purpose of the field. In this case, it can be seen that the field is not brown or yellow, which may cause the auditor to speculate that it is not for growing crops, and so it might reasonably be concluded that it is probably animal grazing land. If this supposition is correct, then the auditor might ask, 'Where are the animals?' The answer in this case would lead us to the fact that the farm was afflicted by 'foot and mouth' disease and that all the animals had been slaughtered. This example should lead us to recognise that the auditor must not only observe what is there but should also be aware of what is missing and what the auditor could reasonably expect to see.

Observations can also be a great help in knowing who to talk to. The next photograph (Figure 10.3) shows a small storeroom in need of some care and attention.

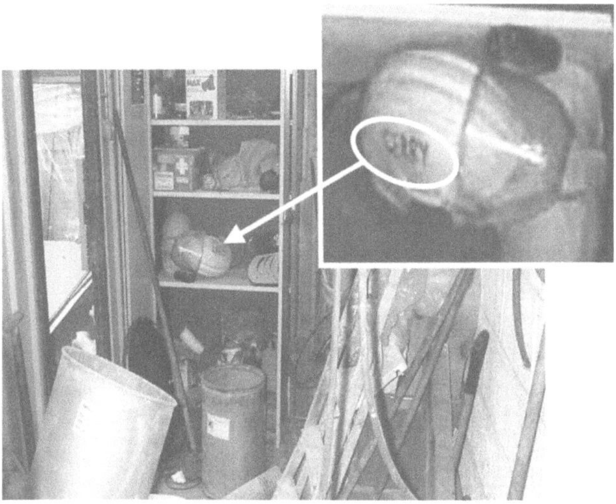

FIGURE 10.3 Auditor's observations: visual clues.

The observant auditor will notice that the hardhat has the word 'Gerry' written on it. If 'Gerry' puts his hard hat in the storeroom, the auditor may reasonably conclude that Gerry may also know something about the state of room.

Audit observations are all part of the detective work of identifying what really happens and play a key part in identifying opportunities for improvement.

It is also important to be precise in what is observed. In the famous example by Frenchman René Magritte, he drew a pipe and wrote beneath it, '*Ceci n'est pas une pipe*' (This is not a pipe; Figure 10.4).

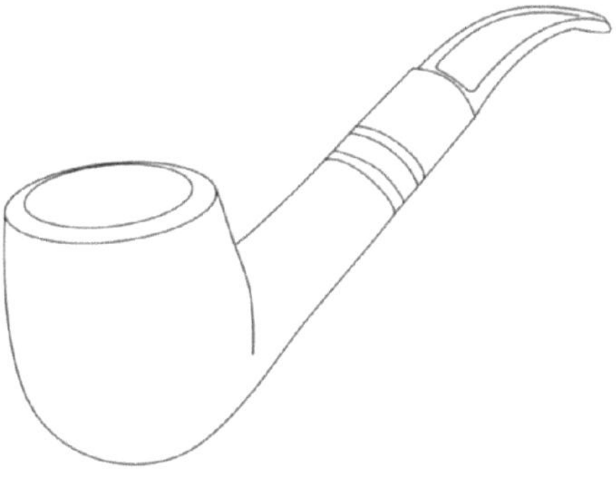

FIGURE 10.4 '*Ceci n'est pas une pipe.*'

He was actually being very pedantic. It is not actually a smokers pipe; it is a picture of a pipe! This precision may be important when auditing as it may be important to know whether you saw what a particular individual did or whether someone else told you what that individual did. Seeing for yourself what is done is a higher standard of evidence compared with hearing about it second-hand. So when we look at procedures, for example, it may be important to know whether we are looking at the current working procedure or a copy of that procedure. In most quality controlled systems, you will find that printouts of procedures are usually labelled as 'uncontrolled copies', as maintaining documentary control of paper copies is much more difficult than controlling electronic versions. The significance of this to the auditor is that if you are provided with pre-printed versions of an uncontrolled paper copy of some instruction, how do you know that it is the most up to date?

Finally, the auditor must be aware of the consequences of direction. Things can appear to be different if they are viewed from different angles. It is quite a common practice that areas of a workplace that are seen and used by the public are very well looked after, but behind those areas the housekeeping may be to a very different standard. The auditor must therefore be prepared to look at things from a range of different points of view to see whether he or she gets different messages. In summary, in order to gain a full picture, the auditor should always look

- At (i.e. directly at the item being studied)
- Above
- Beyond
- Behind
- Beneath

(Also see Appendix A1.13.)

This memory jogger is often known as the 'A^2B^3'.

In summary, it must be remembered that although the observation of evidence is one of the most powerful ways for the auditor to glean accurate information, because it is done by fallible human beings, it can sometimes result in inaccurate perceptions.

11 The Formal Discussion

Without questions, there is no learning.

W. Edwards Deming

The auditees and the auditors are usually very similar people; in fact, my preference is that we change places, so that on one day I may be auditing you and the next day you may be auditing me. Having 'streetwise' part-time but competent auditors is by far the best way of ensuring that a balanced and pragmatic view prevails over the audit recommendations. Unfortunately for the auditor, when it comes to audit discussions people undergo some sort of transformation. The time-served pessimist becomes an eternal optimist. The entire world becomes tinted with rose-coloured spectacles and the manager believes that all his or her standards and instructions are being followed to the letter. This is not a situation of untruths; the competent and committed manager genuinely believes that most things are all right; otherwise, he or she would have done something about it. The task of the auditor during the formal discussion process is to act as a detective and try to separate what actually happens from what the managers and other responsible people believe should happen. The auditor's job is like that of the teacher in the film *Dead Poets Society*; it is to look at the same information as the manager but come at it from a different perspective.

So after the area familiarisation tour, the next major step in the information-gathering process is the formal discussion. Very often, the unit being audited will present its managers to act as the auditees. Consideration should be given to whether managers are always the best people to respond to the auditor's questions. My view is that the auditee should be the most knowledgeable person at the facility in the particular subject being discussed. This may be a manager, but more often than not it could be someone at a less-elevated level in the organisation.

In the past, the audit discussions were called interviews, but every so often the interview developed into an interrogation. The second audit that I ever carried out was at an overseas location. Thankfully, I was not a lead auditor in those days and just watched in horror as the interview process evolved with 5 auditors around a table at one end of this auditorium and 35 members of the site's staff arrayed before us. It was brilliantly 'stage-managed' by the site's audit manager, since it was almost impossible to get sensible discussions going and the whole process seemed more akin to the Nuremberg trial. The following day, we quickly regrouped to adopt 'plan B', which involved the radical concept of one-to-one discussions, and the day was saved.

We have said previously that 'audit' means 'listen', and therefore, the primary role of the auditor is to be a listener; the discussion must not be allowed to turn into an ego trip for the auditor to display how superior his or her knowledge of the subject is compared with that of the auditee. It should be remembered when selecting auditors that the letters in 'LISTEN' can be rearranged to spell 'SILENT'. The Good Lord in his wisdom gave us two ears but only one mouth for very good reasons; so remember that we cannot listen if we are talking. Those people who like the sound of their own voice are unlikely to make good auditors. Listening is an active process and not a passive one. To listen effectively we need to

1. Concentrate on what is being said
2. Show interest
3. Allow the auditee to communicate their full message before replying

Replying is the active part of the process and shows that you have listened and correctly understood what was said.

Generally, the audit discussion should be conducted in the same professional manner as any other meeting. It is worth paying some attention to the ergonomics of the room used. Frequently, the auditor will be provided with a room that is either a disused office or the corner of some enormous conference room. In both cases, take a quick look around before the discussion starts and ensure that there is nothing about the surroundings that conflicts with the intent of the conversation. Ignoring housekeeping hazards such as huge piles of wastepaper in the meeting room effectively means that by default you are approving it and this will compromise your position when commenting on the fire risk associated with similar problems in other people's offices. Always remember that as an auditor *everything that you walk past without comment will be assumed by the auditees to be acceptable*, and so it is essential that you 'walk the talk' and set a personal example.

The layout of the room is important in setting the tone of the discussion. Darkened rooms and angle-poise lamps happily no longer have a role in the process. The emphasis is on setting the auditee at ease and establishing a non-threatening environment. Try to avoid facing each other across the desk in a confrontational manner. If you require writing space and need to sit at a desk or table, sit side by side with the auditee and let him or her see the comments that you are writing. This way you are seen to be sharing the problem and not judging (Figure 11.1).

Open the discussion by explaining the reason you are there, again emphasising that you are not there to judge but to help explain that you are using a checklist/protocol, and ensure that the auditee is happy with that. Wherever possible, encourage

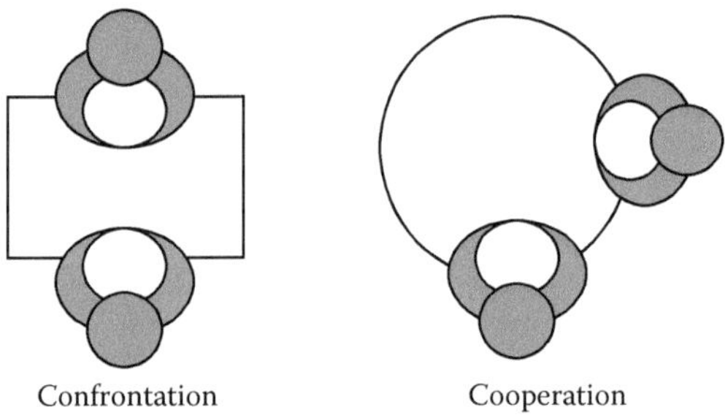

FIGURE 11.1 Discussion arrangements.

the auditee to show you examples of evidence that supports his or her argument and take notes to confirm evidence that has been verified.

Compliance audits that last for only an hour or so and are checking on a specific instruction will probably only have the one auditor present during the discussion. In major SHE audits, which can attempt to cover 100 or so different aspects, from the existence of an SHE policy to the arrangements for the disposal of waste, it is often considered advisable to have two auditors at each discussion. This is to allow one to lead the discussion and another to take notes. My own view is that it is quite possible and very time effective for just one auditor to conduct each discussion, provided that he or she has a fully developed audit protocol and is very familiar with the audit process. If you have to rely on the interpretation of generic checklists during the interview, then the auditor will require more thinking time, and in those circumstances, two auditors may be preferable.

With the discussion programme and room layout/ergonomics sorted out in advance, we come to the event itself: the audit discussion. It is said of student lectures, that the well-accepted principle is for the information to pass from the notebook of the lecturer to the notepad of the student, without it passing through the heads of either. This approach is to be avoided in audit discussions. The principle here is that the auditee be treated as the expert and the auditor as the student. It is the aim in this case for the student (auditor) to gain as much information as possible in order to understand how things are done at this location.

To achieve this, the interpersonal skills of the auditor are paramount. It is worth repeating that the word audit means 'listen', so the auditor must listen and empathise. There is no place here for finding fault, ridiculing, punishing or blaming. Non-verbal signs and tone of voice account for the greatest part of any communication, so avoid showing signs of frustration, shock or annoyance (Figure 11.2).

Dr Albert Morabian's research tells us that only 7% of our message is the words that we use, 38% is the way we use those words and 55% is the body language that we adopt.

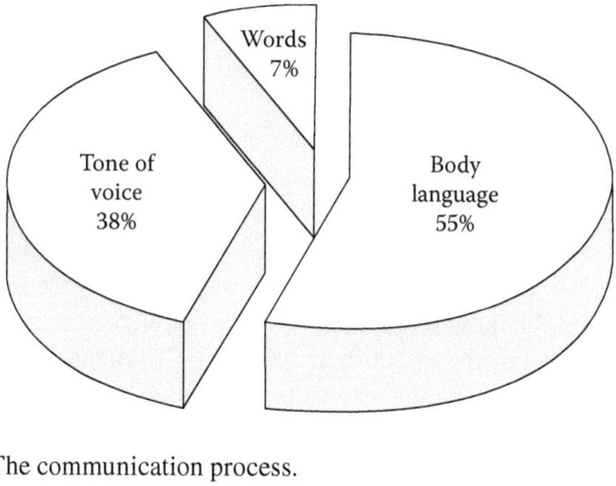

FIGURE 11.2 The communication process.

This means that the auditor will have to carefully and unobtrusively manage the time allotted to each discussion, as the most common cause of auditor frustration is to overrun the time allocated for discussion.

The prerequisites for the discussion, in addition to the hygiene factors of room layout and ergonomics, need to create an atmosphere where both parties in the audit discussion have an incentive to participate. The auditor clearly has an incentive, but the auditee may need some encouragement, as he or she will have arrived concerned about being criticised or fearful of a lot of extra work. To achieve this, the auditor must have credibility and gravitas. An auditee with 35 years' experience in the job will not take kindly to being told what to do by an auditor straight from college. If the auditor is not known to the auditee, it is useful for a brief CV or résumé to have been sent in advance or for the auditor to have summarised his or her relevant experience at the entry meeting.

The discussion should have a clearly stated purpose, and so it is helpful for the auditor, after the essential preliminary small talk and pleasantries, to briefly explain the process and the fact that the overriding purpose is to contribute to further continuous improvement in the organisation's SHE performance and to try and ensure that neither people nor the environment are harmed.

Although the auditor will be prepared with the prompt questions on his or her checklist or protocol, these are not normally used at this early stage of the discussion. The idea is to get the auditee talking about his or her particular aspect of the audit and to elicit answers to the pre-prepared questions out of the discussion. This approach sustains the discussion and avoids it becoming an examination. A good opening gambit is to use these words:

'Please help me understand how you manage …'

My experience is that with some judicious steering of the conversation, this approach will answer 70% of your pre-prepared checklist questions.

Take the opportunity during the discussion to give personal recognition for points of excellence using such phrases as

'That's a great idea.'
'If it's OK with you, I'd like to mention your solution to … as I know they have a similar problem.'

When giving recognition, always pause for a second or two before moving on to allow the point to sink in. Remember that it is very easy to undermine recognition by the use of the word 'but'. For example, saying, 'I really like what you have done here, *but* why didn't you do it this way …' infers that you are cleverer than the auditee and not only undermines the recognition but also undermines your relationship. If you are successful in establishing a relaxed conversation, then you will quickly find that people actually enjoy talking about their work. The challenge then becomes one of reining them in without losing their or your own enthusiasm or putting them down.

Occasionally, when you pose a specific question, the answer *'I don't know'* will come back. This is actually encouraging, because the auditee is sufficiently relaxed to know that he or she doesn't need to *'fob you off'*. However, you still need an answer, so don't lose the opportunity to ask, *'Can you tell me who might know?'*

Encouraging further information can be done by the use of such phrases as

'Go on.'
'Can you give me some more detail about …'
'That's very interesting, can you explain to me how you do …'

Silence can also be used very effectively to elicit more information. Often during the conversation, the auditee may drift from the first-person singular ('I do this …') to the first-person plural ('We do this…'). Generally as auditors, we are interested in what individuals do, because the use of the word 'we' infers what should be done, rather than what is actually done. So if the word 'we' is repeatedly used, then bring the discussion back by saying something like

'You say "we", but what do you actually do?'

Of course, all that part of the discussion is the easy bit. From time to time, it is inevitable that the auditee will either tell you something or will omit something that implies to you that he or she doesn't comply with the required standard. I have heard auditors say things like

'I can't believe you do that.'
'Did you know that you could go to jail for that?'
'That was a stupid thing to do.'

All these and a multitude of other inappropriate responses purely cause auditees to retreat into themselves and become defensive. So if it becomes obvious that something isn't right, it is best to get the auditee to admit/identify what is wrong for themselves, rather than for you to tell them. Examples of phrases that can be used are

'Do you think that is right?'
'Is there anything else that could be done to prevent …?'
'You've said that you see a problem here; what would you do about it?'

(More possible audit questions can be found in Appendices A1.11 and A1.12.)

The auditor can also gain useful information by the way in which questions are answered. The use of the words 'would', 'should' and 'could' in replies almost invariably infers a lack of certainty in the answer. For example:

'I would have thought that …' *really means* 'I haven't a clue!'

or

'I should do it like this …' *really means* 'I know how it ought to be done, but I do it a different way'.

whereas

'I could have done it that way...' *really means* 'That is another option, but I don't think it is a very good one!'

During the discussions, it will be necessary for the auditor(s) to take notes. It is important to keep an accurate record, and although it will have been mentioned to the auditee at the beginning of the discussion, the actual note-taking should be done as unobtrusively as possible. The use of computerised note-taking is often more threatening to auditees than handwritten note-taking, and using tape recorders or other voice-recording devices has a feel of clandestine surveillance and should be avoided.

Once you have gained all the information that you need or all that is available, then the discussion needs to be closed. This is the area where the audit discussion differs from most normal conversations. The auditor needs to ensure that he or she has gained a correct understanding of the discussion. This is done by a short summary of the key actions that have arisen, particularly where these are likely to result in some sort of corrective action. This gives the auditee the opportunity either to agree with the summary or to correct a misunderstanding. It is this part of the audit discussion that differentiates it from other conversations and is the thing that is most commonly overlooked by inexperienced auditors. Finally, the discussion is closed by thanking the auditee for their time and re-emphasising any points of positive recognition.

All the members of the audit team in any one audit should adopt the same conventions of note-taking during discussions in order to streamline the later stages of verification and report writing. It is in this standardisation of discussion-recording convention that the use of a protocol scores strongly over simple generic checklists. A correctly designed protocol will not only guide the auditor's questioning but will also prompt the auditor when evidence or verification will be required. Usually, the auditor will write the interview notes in pencil, as this will allow him or her to return and amend the records as more information becomes available during the audit. The usual convention when the auditee's comments have been confirmed through other evidence is to annotate the comment with a letter 'V', signifying that it is not just hearsay but has been independently verified. Matters that arise during the discussion that require further follow-up or verification would normally be highlighted using a brightly coloured highlighter marker pen so that these do not get missed later in the process. Highlighted subjects, which are subsequently verified as acceptable, are then marked with a 'V' in the normal way.

It is also advisable to have annotations for highlighting significant noncompliances and matters of excellence that will form part of the positive recognition outcome of the audit. My own convention for these is to annotate them with 'R' (for 'potential recommendation') and 'E' for 'area of excellence'. Conventions for annotating audit notes:

Area of excellence E

Potential recommendation Ⓡ

Verified Ⓥ

Noncompliance Ⓝ

Major audits, which attempt to cover a wide spectrum of aspects of SHE issues, could require a large number of separate discussions. It must be recognised that the auditor has the opportunity to create total chaos in the normal smooth running of the unit by placing unreasonable demands on auditees by repeatedly calling the auditees back at different times to discuss different subjects. Wherever possible, the discussions should be grouped to minimise the disruption on the auditees rather than to slavishly follow the order in which the subjects appear in the local SHE manuals. If Bob Smith (say) is nominated to discuss standard number 1 (Safety Policy), standard 6 (Safe Systems of Work) and standard 20 (Waste Disposal), then don't call him back three times for the three different subjects but arrange a discussion session with Bob Smith that deals with all three subjects in the one visit. The only thing to remember is to ensure that Bob Smith knows in advance that he will be expected to comment on all three areas. The general comment is that formal discussion programmes should be set up on the basis of people's availability, not just in the arbitrary order that the subjects are listed in some manual or other. It is advisable to plan periodic 15-minute breaks or dead time into the programme of a major SHE audit, since if the auditors are involved in continuous discussions all day, there is a risk of overrun on some subjects and there is nothing worse for an auditee than to be kept waiting. The presence of the dead time can act as a buffer to prevent major overruns. Although the purpose of the audit discussion is one of listening and gathering information, the auditor must keep control of the formal discussion debate. Frequently, it will be found that people like talking about things that they do well. They may have a desire to go into necessary detail when the auditor has already concluded that they are in compliance with the requirement. Discussion control will be a balance between the auditor getting the information that is required and auditees feeling that they have had a fair opportunity to explain how their system works. On the other hand, if it becomes obvious that the site is not compliant with some particular aspect of the audit, do not persist with the checklist or protocol questions on that subject relentlessly and end up embarrassing the auditee to such an extent that he or she becomes reticent. At the end of each discussion, give some feedback to the auditee, particularly about the areas of excellence, areas where further verification may be required and areas where there may be suggestions of noncompliance. If possible, seek his or her agreement, particularly if there are areas of possible noncompliance, because although it is important not to jump to conclusions at this early stage in the audit, it is important that the auditee and the local management get a feel for where the auditor is starting to see areas for comment.

Formal discussions are an intense and exhausting activity for both auditors and auditees alike. In a major audit that lasts several days, it is worth splitting these up into two or more blocks to allow some variety and breaks for all concerned (Figure 11.3). This approach also allows the auditor to carry out some verification actions while they are still fresh in his or her mind, before embarking on developing yet more lists of items for verification.

	Day 1	**Day 2**	**Day 3**	**Day 4**	**Day 5**
Morning	Entry meeting and site tour	Formal discussions	Verification, practical inspection, and informal discussions	Formal discussions	Verification/ inspection/ informal discussion
Afternoon	Formal discussions			Verification/ inspection/ informal discussion	Auditor meeting and Exit meeting

FIGURE 11.3 Example of spreading the formal discussion workload on a major management audit.

Auditors are sometimes anxious about asking obvious or apparently stupid questions. You will not be expected to have a detailed understanding of the technology, and often, local personnel don't ever ask themselves some of the basic questions. It should be remembered that often the only silly question is the one that you didn't ask.

(For summaries of formal discussion guidance, see Appendices A1.10, A1.11 and A1.12.)

12 The Informal Discussion

Too much agreement kills the chat.

John Jay Chapman

The formal discussion process plays a major role in helping the auditor understand what should be happening in terms of SHE management in the organisation. Unfortunately, what should be happening and what actually happens are not always the same thing. To find out what actually happens, it is necessary to talk to a cross-section of employees and also to observe their actions and behaviour. The main feature of informal discussions is that they are often unplanned and opportunistic. They should in fact be conducted as fairly casual conversations, like having a chat. The informal discussion will often take place while the auditor is conducting either an area inspection or verification activities. While the auditor is out and about, he or she will need to take the opportunity to listen to as many of the employees or other stakeholders as possible to try to get an understanding of how standards are actually applied. Whenever possible, the auditor should carry out the informal discussion in the employee's normal work area; here the employee is on home ground and is less likely to be intimidated by the concept of being caught by an auditor. When approaching people in the workplace, auditors should always introduce themselves and explain the reason they are there. The auditor should indicate interest in helping ensure that no one comes to any harm and that the organisation is in compliance with the required standards. The conversation should then move to discussing aspects of safety, health and environmental performance. The auditor may be interested in something that the employee was or was not doing, or the auditor may wish to follow up some particular verification action identified in the earlier formal discussions, or he or she may wish to test the employee's understanding of one or more of the organisation's standards or instructions. In the informal discussion process, the emphasis will not be so much on 'Tell me how this or that happens', but rather it should focus on 'Show me how it happens' or where certain information or records are kept. The sorts of questions that may arise are

- 'Why are you doing it that way?'
- 'When did management last discuss SHE with you and what did you talk about?'
- 'What training have you had to ensure that you understand the risks of your job?'
- 'Why shouldn't that liquid be spilled on the floor?'
- 'How do you make sure that you cannot be harmed by this task?'
- 'Show me where I can find copies of the health and safety instructions.'
- 'Show me how you would isolate that equipment.'

- 'Show me what you would do if the fire alarm sounds.'
- 'Show me what protective equipment you use to do this job.'

The importance of using the 'show' rather than 'tell' approach is crucially important in aiding verification. 'Showing' entails the use of verbal and visual information concurrently. This combined verbal and visual confirmation of information means that facts provided in this manner may require no further verification.

The informal discussion will not usually follow a predefined set of questions. If the auditor needs to be shown a local health and safety instruction, instead of searching for it him- or herself, the employee should be asked to find that particular instruction. This can then lead to talking about what training the employee has received in that instruction and how he or she was validated. In this way, using an informal but logical flow, the auditor not only establishes whether the instruction exists, but they will also be able to tell by its condition whether it is well thumbed and used or whether the pages still crackle when opened, signifying that it is rarely opened. The discussion about training can help establish whether the individual has been trained, and if not, the auditor will then need to establish whether the lack of training is a unique omission in the case of that individual or the norm for all employees carrying out that task. The selection of people at random is an important element of the informal discussions. The auditor will need to set a target for covering a reasonable proportion of the employee population so that the results can have statistical significance (see Chapter 13). Before closing the informal interview, the auditor should thank the employee for his or her help and give the individual the opportunity to mention any safety, health or environmental concerns that he or she thinks may be important that the auditor should be made aware of.

Most of the discussion principles of the informal discussion are the same as for the formal discussions, but because the meetings will be unplanned, a simple checklist of the discussion process can be summarised as follows:

1. Explain who you are and why you are there.
2. Confirm that it's OK; if not, arrange a more convenient time.
3. Allow the other person to do most of the talking.
4. Use 'open' questions (i.e. questions that don't just have a yes/no answer).
5. Do not use 'leading questions' or questions that bias the answer.
6. Test your understanding of what the person is saying.
7. Remember to listen and seek amplification.
8. Before leaving summarise your understanding.
9. Thank the person for his or her time.

The informal discussion is a very important part of what is known as 'drill-down'. This is the process that the auditor uses to examine certain aspects of the audit scope in more detail in order to get a clearer idea of what actually happens and hence verify or contradict what has been previously said in the formal discussions. During auditor training, which should always involve some practical experience of auditing on-site,

I find that trainee auditors tend to be reluctant to just stop people in passing and have a chat. On large sites, such as chemical works, construction sites, farms and utility operations, there may not be very many people about and so I recommend that the auditor should take every opportunity to talk people that they meet informally. It will become obvious very quickly if they have anything to say that contributes to the audit.

13 Statistical Significance

It must be remembered that any auditing process is based on sampling what happens, both in terms of activities and documents. No audit can achieve a 100% representation of reality; at best, it is taking a snapshot in time. It is therefore crucially important that the snapshot be as representative as possible of what really goes on in the organisation. Clearly, a single discussion with the chief executive of a large car manufacturing plant will not necessarily give a truly representative view of what happens on the shop floor. The real question is, What level of audit examination could give a reasonable chance of getting a reliably representative picture of what really happens?

Many of the commercially available auditing systems provide a quantitative assessment based on a wide and diverse range of questions. Although the questions themselves are very relevant, the systems may fall down at the user level. This is because the user and the person who completes the assessment is typically a manager or group of managers within the location being audited. More often than not, in my experience, the audit is carried out almost entirely (and sometimes only) by the organisation's safety or environmental manager, who may find that his or her personal performance rating depends on the outcome. Often, these assessments are carried out in an office without any involvement from the other employees. It must be remembered, therefore, that these audit results actually represent a management view of the situation and are likely to be somewhat optimistic and one-sided. The most essential part of any audit is the verification and opinion sampling step, and this is the step that is so often missing from some of the 'self-audit' processes. It is this activity and this activity alone which has the ability to transform the management view into one that more closely aligns with what actually happens. It is a statement of the obvious to say that the quality of any sampling process will affect the quality of the output of that process. A sample of one employee in a population of ten will be more likely to give a representative result than taking a sample of one employee in five hundred. To give credibility to their conclusions the auditors should set themselves targets of the number of people that they are aiming to talk to in the organisation. Included in these figures will be the discussions that take place both in the formal and informal parts of the audit process. Figure 13.1 gives an indication of the approximate number of personal discussions that an audit needs to achieve for the results to have statistical significance for given sizes of population.

In general terms, the larger the number of people employed at a facility, the more people the auditor will need to talk to. However, at small locations, the auditor will need to talk to a *higher proportion* of the staff.

However, the auditor must always recognise that any sampling process could infer that there is some degree of uncertainty in the results and he or she must factor this uncertainty into the audit conclusions. A conclusion that has profound consequences for the organisation but is based on a very small sample size may not be robust. In these circumstances, it will be necessary to do further sampling by doing some

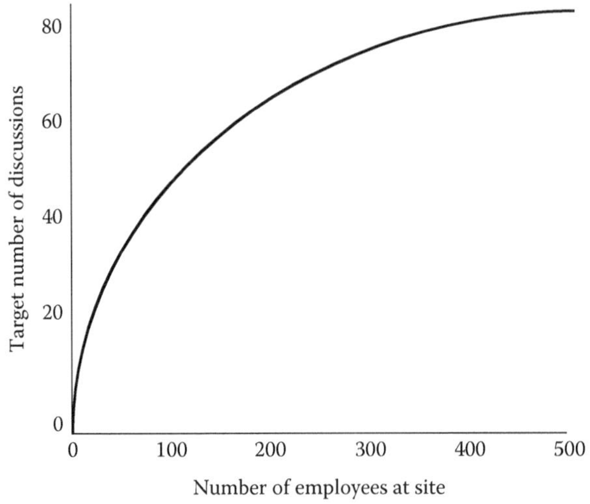

FIGURE 13.1 Discussion targets vs. site population.

more detailed 'drilling down' to increase the sample size and therefore the amount of evidence available to support or challenge the auditor's conclusion. During the reporting stage of the audit, it is always advisable to record and report the number of discussions that have taken place as a proportion of the location population.

When it comes to reporting, it can be helpful to give an indication of data statistical reliability. So if you are making an audit recommendation to improve the maintenance of fire extinguishers, it can be helpful to indicate that 6 out of 25 fire extinguishers checked were found to be overdue on their annual maintenance.

14 The Importance of Verification and the Audit Trail

Evidence is often defined as 'a thing or set of things helpful in forming a conclusion or judgement'.

If the audit recommendations are to have any credibility, the auditors will have to provide such evidence to support their conclusions; without this evidence, the audit process and any recommendations are based on 'gut feel' and are not likely to be taken seriously. In a court of law, it is only corroborated evidence (i.e. evidence substantiated from more than one source) which is irrefutable. Likewise, in SHE auditing, every effort should be made to seek sufficient evidence to support any claims that are made. Any recommendations that are made following the audit must be supported by facts. Otherwise, the audit will be inaccurate and the auditor's credibility will be undermined. It is a common new auditor's mistake to find one piece of information that contradicts some local requirement and then immediately assume that this is representative and rush into a recommendation. Audit evidence consists of the information that the auditor uses during an audit to substantiate the audit conclusions. The auditor will encounter many different types of evidence (written, verbal, observations). To properly evaluate the strength of evidence, the auditor must understand the four concepts of evidence:

- *Format*: The form of the evidence – for example, verbal, visual or written (see Chapters 10, 11, 12 and 16)
- *Suitability*: The quality, relevancy and reliability of the evidence
- *Sufficiency*: The quantity of audit evidence – enough evidence to evaluate the audit client's management assertions
- *Evaluation*: A decision on whether the evidence is sufficiently compelling to allow the auditor to form an opinion

This process of seeking confirmation of evidence is usually referred to as the 'verification process'. This process of verification is part of what is known as 'drill-down'. This is the colloquial expression for the approach auditors use to dig deeper and deeper into the organisation's systems in order to find out what really happens. Verified evidence may come in the form of corroborative verbal statements, visual observations of situations and behaviours or alternatively through documentary evidence. Evidence that has been independently sought by the auditor, rather than voluntarily provided by the auditee, is the most valuable, since this is least likely to be biased or tainted. The need for verification is usually first identified in the formal discussion process. You will recall that we identified the need to highlight all the potential areas for verification in the auditor's working notes or protocol. At the end of the discussion processes, the auditor(s) will have a very large number of topics that need verification. For an audit with more than one auditor, the lead auditor needs to manage the team carefully at this stage to ensure that the verification process is both efficient and avoids the different auditors making a nuisance of themselves by repeatedly going back to the same people for different information. A good example of this is when the audit team is reviewing training records. Every SHE audit will identify a need to verify the existence of training records for a range of different subjects. The auditor(s) should endeavour to collate all the verification requirements for training records and make one visit to the training record holder to deal with them all. The only issue for the auditor(s) is how to quickly and efficiently collate their verification requirements into logical groupings. My own experience is that it is quite normal to have 200 or more requirements for verification on a major SHE audit and dealing with these in an efficient manner is no trivial matter. The most effective way to do this is by transferring all the verification requirements one by one onto separate sticky notes. With this method, the verification requirements can easily be grouped and regrouped to allow them to be allocated to particular auditors or linked to particular employees. When the auditor goes to carry out the verification, he merely collects the relevant sticky notes to use as a memory jogger (Figure 14.1).

Sticky notes that have been satisfactorily verified from more than one source are then annotated with the 'V' convention to signify that they have been verified and are parked in a completed file or location. Sticky notes that cannot be verified are at this stage starting to be recognised as possible issues and may require even more focused and detailed attention. These sticky notes will be displayed on an issues board or what we colloquially call the 'sin bin'. The delight of this method of controlling the verification process is that it gives the auditors an immediate and visual picture of the progress that they are making, because points for verification that are still stuck to the wall or tabletop are still outstanding work for the auditors. More importantly, it gives the local management team a visual and real-time display of what the auditors are finding (Figure 14.2).

This latter point is particularly important from two points of view. First, it allows the local management team to challenge the auditors' understanding of what they have found, and it is much better for the auditors' credibility to find that they have misunderstood someone's comment before they start building houses on faulty foundations. Second, it allows the local management to start to recognise issues before they are sprung on them at the exit meeting. This ongoing method of feedback is greatly valued by auditees and auditors alike.

The sticky notes can also be used to identify the points of excellence on an ongoing basis as well as points for verification and issues. Although these may not require further feedback or verification in the way that other points do, it is politically astute to display them in what we term the 'grin bin'. It is worth working hard to ensure that the grin bin contains a reasonable number of sticky notes.

Most verification actions are followed up through the informal discussions described earlier or via documentary review. However, on occasion, it may be necessary to get views on how effectively certain things are carried out throughout the

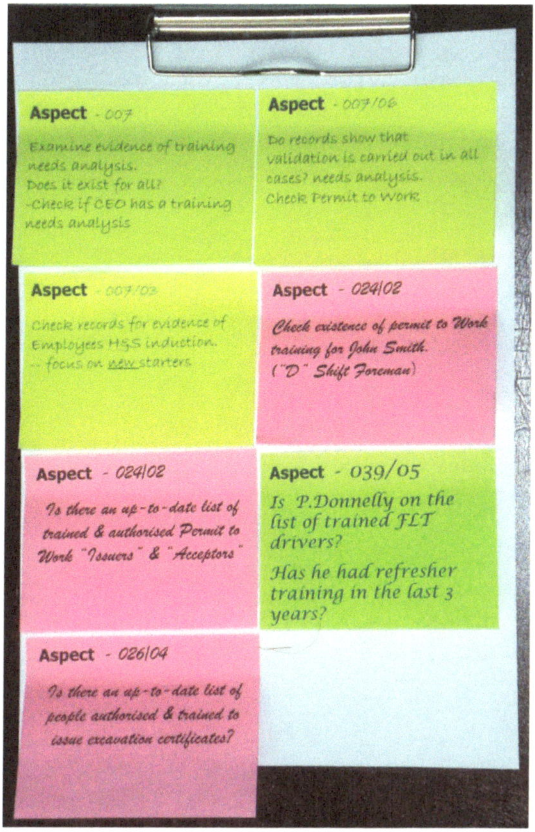

FIGURE 14.1 Example of grouped sticky notes on a clipboard. Different coloured sticky notes have come from different auditor's formal discussions.

FIGURE 14.2 An auditor reviews sticky notes with the audit manager.

whole organisation. In this case, it is sometimes useful to prepare a simple questionnaire that all the auditors can use to get a quick but wide sample. For single-auditor audits, this technique can be used to get wide rapid feedback by delegating the task of getting feedback to the questionnaire to a local manager. It is unlikely that the questionnaire can be fully prepared in advance, as the questions posed may need to reflect some of the potential issues identified by the auditor (Figure 14.3).

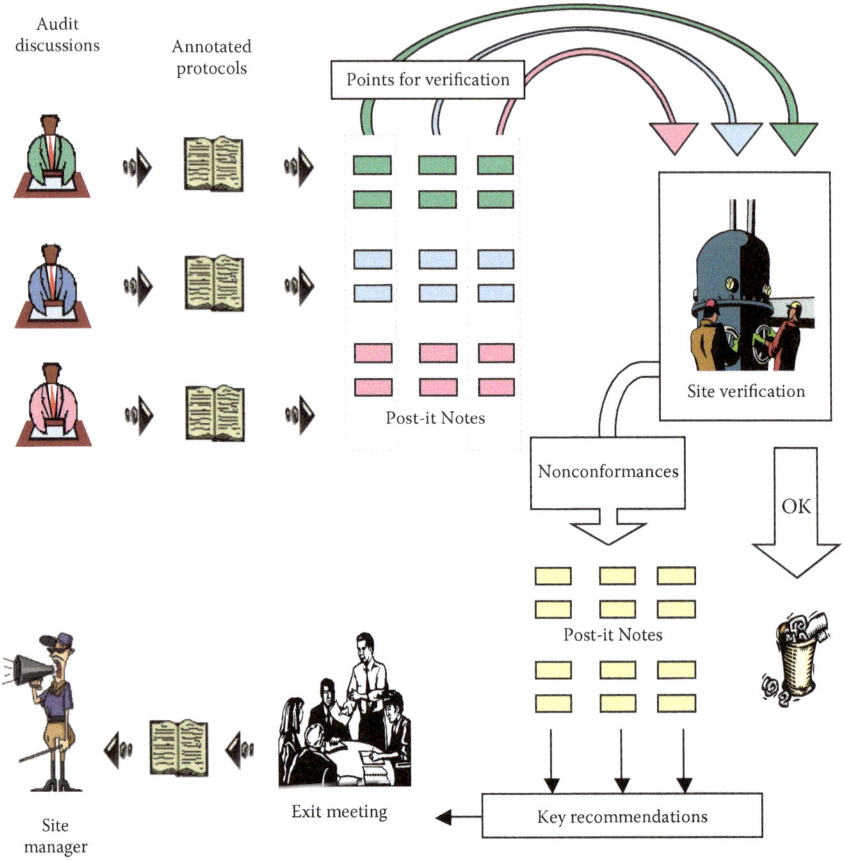

FIGURE 14.3 The Plaudit 2 process for health and safety audit analysis.

What happens if there is no local requirement for a standard or instruction when the auditor clearly observes a need? This situation is not uncommon, especially in well-run smaller organisations, and quite often reflects a situation where the local requirement has never been written down. The fact that the specific failing is that 'the detailed requirements have not been documented' makes it more difficult to audit, but it does not necessarily mean that nothing is happening; there may well be perfectly adequate practices in place. The concern might be whether that practice can stand the test of time without being written down. In this case, the verification process has an essential role to play in establishing what level of practice exists and how it might compare with the auditor's understanding of best practice. However, I repeat that the danger in these situations is that the practice usually relies on the knowledge of a very small number of individuals. Once that person leaves or is absent for a while, then the good practice can very quickly deteriorate into no practice at all. In these circumstances, where good practice is in place but with no documented requirement, the auditor is likely to make a recommendation that 'the organisation's good practice needs to be incorporated into a formal procedure or instruction'.

15 Conformity

Audits are the most wonderful learning events. Not only does the audited unit learn where it has opportunities for improvement, but audits are also a learning opportunity for auditors. This is why I believe so strongly in not using full-time professional auditors for the purpose, as they are rarely in the position to make best use of the learning. Practising and experienced line managers with a good knowledge of SHE requirements will have the highest credibility. The auditor's learning not only applies to major audits at the management and specialist level but also applies equally and probably more so to the compliance-level audit. Managers all too rarely sit down and learn their own safety instructions, particularly if they have inherited them rather than written the instructions themselves. There is no substitute for auditing someone else's compliance with an instruction or requirement to test one's own understanding. There is no doubt in my mind that involvement in an audit process is one of the best management training techniques.

Since formal quality processes were first introduced in the latter part of the last century, adherence to the specified requirements was usually referred to as 'conformance' and deviations from the standard requirement were usually termed 'noncompliances' (or sometimes 'nonconformances') and form the essence of the audit feedback. For many years, these terms were the de facto standard way of referring to how organisations met their requirements. The terms 'conformance' and 'compliance' were generally used interchangeably, and you will have noticed that thus far in this book these are the terms that have been used because they are still by far the most commonly used terms in this respect. However, in recent years, the development of the International Standards (ISO 9000 series and ISO 14001) have standardised terminology. In order to differentiate between regulatory requirements and internal standard requirements, the terminology has been standardised so that 'compliance' and 'noncompliance' are reserved for descriptions relating to regulatory issues, and the terms 'conformance' and 'nonconformance' relate to nonregulatory requirements. Unfortunately, the standardisation of terminology does not end there. Even more recently, the word 'conformance' in all the ISO standards has been superseded by the new term 'conformity'. It is expected that this new term will progressively become the norm, and so from now on, throughout the remainder of this book, I shall refer to the terms 'conformity' and 'nonconformity' in relation to adherence to nonregulatory requirements.

From here onwards, we shall use the International Standards terms

- **'Compliance/noncompliance' to relate to regulatory requirements**
- **'Conformity/nonconformity' to relate to nonregulatory requirements**

Often, the audit will identify a wealth of information, and much of that will form the basis for the audit feedback and recommendations, some of which will be directly related to the scope of the audit and some of which may be outside the scope or the particular requirement being assessed.

However, there may be other things that the auditor sees, where he or she believes there is some learning but it does not form part of the audit requirement. Typically, if an auditor is carrying out a specialist occupational health audit and notes during the site inspection some potential trip hazards, it would be irresponsible of the auditor to ignore those hazards. Equally, if the auditor is auditing fire safety requirements and sees that the local requirement does not call for routine testing of the fire pumps, even though it is not a documented requirement, the auditor may wish to pass comment about the advisability of introducing such tests. In both these cases, the local requirement has been met, so strictly there is no nonconformity, but the auditor may choose to make learning *observations* in the audit report. Provided that the number of observations does not get out of hand, the auditor should feel free to differentiate between nonconformities and observations. After all, there will be few occasions where the SHE requirements will be studied more closely than during an audit, so every effort should be made to extract the maximum amount of learning. Equally, it must be remembered that making a multitude of irrelevant and unhelpful observations will not endear the auditor to the auditees. It must be remembered that it is not the role of the auditor to show how clever he or she is! Before incorporating 'observation' recommendations within an audit report, always check with the auditee's senior manager to see if these will be welcomed, as they are actually outside the agreed scope of the audit. My experience is that these observations are nearly always welcomed. If some organisations are resistant to reacting to auditors 'observations', one of my clients helpfully refers to these recommendations that are outside the agreed scope of the audit as 'risk reduction measures'. It is difficult for any management team to argue against anything that results in risk reduction!

Very often, observations will be made in relation to shortcomings in the procedures themselves. The audited unit may be in full conformity with the local procedure, but the auditor may have noticed that the procedure has not been updated to reflect current best practice or a recent change in regulatory requirements. In this situation, it is entirely appropriate and helpful that the auditor comments, *'Although in compliance with the procedure, the location is not in conformity with regulatory obligations, and therefore, the procedure needs to be revised to reflect current regulatory requirements.'*

Whenever nonconformities or significant observations are made, the auditor should discuss these with whoever is involved at the time. It is not the auditor's role to go sneaking around in the dark in some clandestine way, furtively noting down failings that can then be used in some dramatic revelation later. The credibility of the auditor and the audit process is dependent to a large extent on the auditor's openness.

Care must be taken when interpreting audit nonconformities into recommendations. The auditor must remember the authority vested in him or her and must be careful not to abuse that power. It is very easy to set hares running in all directions or to commit the organisation to unnecessary cost or impossible tasks. In one audit on a chemical plant, the auditor was checking that the plant identification numbers complied with the local procedure. One of the criteria for numbering equipment related to the situation where two or more heat exchangers are stacked vertically above one another. In this situation, the procedural guidance was as shown in Figure 15.1a, where the lower exchanger should carry the suffix 'A' and the upper exchanger should carry the suffix 'B'.

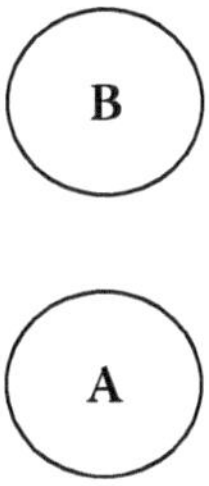

(a)

FIGURE 15.1a Required heat exchanger layout.

One of the actual exchanger sets is shown in Figure 15.1b. The auditor noticed that the exchanger set labelled C1727 actually had the 'A' exchanger at the top and the 'B' exchanger underneath, which was contrary to the local guidance. During maintenance and cleaning, the end covers that carry the identification number have to be removed to allow the tubes to be cleaned by high-pressure water jets. The auditor's first thought was that during a previous maintenance task, the two end covers had inadvertently become switched and replaced incorrectly.

(b)

FIGURE 15.1b Actual heat exchanger layout.

The auditor checked the equipment layout drawings and established that there had been no maintenance error and that the exchangers had been installed in line with the original design. Now there was a problem, as the auditor had established that two major plant items did not conform to the plant procedure. What should the auditor do? The easy solution might be to repaint the end covers to show the 'A' exchanger below and the 'B' exchanger above. However, before making apparently 'obvious' recommendations, the auditor must always consider the consequences of his or her recommendation. In this case, because the original nonconformity occurred at the design stage, simply switching the numbers painted in situ would mean that the exchangers would no longer match with the design drawings, piping and instrument (P&I) diagrams, maintenance histories or manufacturer's serial numbers. To change all of these records would be very expensive and runs a very high risk of some hidden records remaining unchanged and resurfacing at some stage in the future. Furthermore, the plant was more than 25 years old, and many of the operators had worked on the plant ever since it had been commissioned. The operators all knew and expected that the C1727A exchanger would be above C1727B. Simply changing the numbers would not only cause huge problems with the records but could confuse the operators, and in an emergency this could lead to a serious safety hazard. In this case, the auditor recognised that this requirement was only 'guidance' and so noted in the report, 'The C1727 exchangers do not conform to the guidance in the procedure, but in this case, *because of the unacceptable consequences*, no changes are recommended, other than communicating the fact to the plant personnel.' By doing this, the auditor showed that he was being observant, but that he was also aware of the consequences of his recommendations. He had shown that he had not only considered the risk of maintaining the 'status quo' but had also considered the risk of that change. It must never be forgotten that SHE auditing requires judgement by the auditor. Auditors who cannot use their judgement and experience in recognising the scale of their recommendations are unlikely to be viewed as helpful or competent.

16 Documentary Review

If a man will begin with certainties, he shall end in doubts; but if he will be content to begin with doubts, he shall end in certainties.

Sir Francis Bacon
The Advancement of Learning

Sir Francis Bacon has some useful advice for SHE auditors. If auditors begin with preconceptions and biases, the audit will be worthless, but if they progress from a position of healthy scepticism, then their conclusions will be robust. The key method that effective auditors use to move their doubts to certainties is through the processes of 'drill-down' and verification. As we have seen previously, much of the auditor's initial information arises from the spoken word at formal or informal discussions. Verification is the confirmation of this initial information by other supporting information that may be other verbal comments, the auditor's own visual observations or most often from some form of written or documentary information. This process of verification or 'drill-down', as it is often known, is a fundamental part of obtaining reliable audit conclusions and is the place where many inexperienced auditors flounder. The situation is relatively easy in a Level 1 compliance audit, which may be looking at compliance with a single procedure or instruction. In these circumstances, the auditor needs a thorough understanding of the procedure and the limited associated paperwork. The verification still needs to be done on a sampling basis, but sampling done intelligently by the auditor can ensure that he or she is not just directed towards a good outcome. For example, if the Level 1 audit is assessing exhaust ventilation arrangements to ensure that workers are not exposed to harmful fumes, the auditor might take note of one or two particular exhaust systems during the site visit and then, instead of viewing all the test data for every extract unit at the facility, specifically ask to be shown the latest test data for the one or two extracts that he or she picked at random. Alternatively, the auditor might initially ask to see any 'blacklist' of overdue tests on extract systems and then explore why those items are overdue. It is this type of prior thought about how the audit process may be streamlined that makes it both efficient and effective.

The biggest single problem with using documentary information for verification in a Level 3 management audit is the sheer scale of the task. Even in a small organisation, the amount of paper and computer records and instructions could take an army of auditors months to read and digest in its entirety. It is here that the auditor's skill and judgement play an important role in making the task sensible.

Here I advocate the use of the RCRC hierarchy (Figure 16.1), so called because it helps me remember the four steps (reason, choose, read, challenge) in simplifying the document review process.

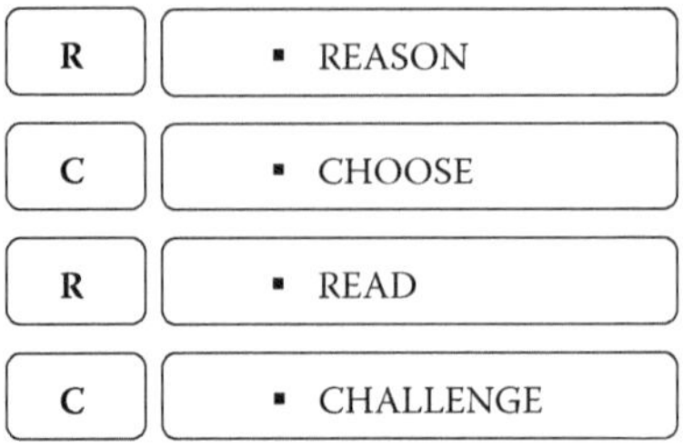

FIGURE 16.1 The RCRC hierarchy.

STEP 1: REASON

Identify the reason you need to consider the particular documents. Ask yourself:

a. Are these documents relevant to the subject?
b. Are the documents current (or have they been superseded)?
c. Has this aspect already been verified (i.e. no further drill-down is necessary)?
d. Does it matter (i.e. the auditee is substantially noncompliant and further examination will not help)?
e. Is this an important issue or a trivial point?
f. Is this something that, because of your special expertise, you already know the detailed requirements for?

If there is a reason to carry out a more detailed documentary review, then proceed to the next stage in the hierarchy ('choose').

STEP 2: CHOOSE

Unless there is difficulty in establishing whether or not conformity exists, we need to select the documents/information that will most easily bring us to a conclusion regarding conformity. So ask yourself:

a. Is the information available? (Don't spend hours looking for information that doesn't exist.)
b. Is the information concise? (Look initially for summaries, flow diagrams and action lists.)
c. Has the information been referred to in discussions? (Ask to see that documentation at the time.)
d. Can a discussion with a different person more easily verify the point?

Experience has shown that there are certain types of information held by most organisations that are almost always viewed during the documentary evidence stage of verification. These include such things as

- Training records
- Operating procedure indices

- Periodic inspection reports
- Routine regulatory submissions
- Internal audit reports
- Relevant maintenance records
- Accident and incident records
- Emissions permits

Again, being selective can be very helpful. If the formal and informal discussions have not identified where training nonconformities exist, then ask about those people who are most likely to have needed training recently. These may be new starters or interdepartmental transfers and promotions. However, always make sure to ask not only about training completed but also how that compares with the training needs and requirements. If there is still a reason to carry out a more detailed documentary review, then proceed to the next stage in the hierarchy ('read').

STEP 3: READ

If we have identified a reason for a detailed documentary review and have selected the relevant documents or databases, then there is nothing for it but to get on and read it in full to assimilate the information. However, there are still some things that we can do to make the process easier for ourselves.

If the document or database has an index or contents list, read that first; you may find that the key information to verify a particular point is actually in the last appendix. Try going straight to what you think is the relevant part of the document that will confirm the point that has arisen for verification.

Scan reading can also be helpful. Try reading the following paragraph:

> Alhtough we may simtmes dobut it, sestyms are crateed to spmlifiy acevititis taht are retpeetad and are esiasntel to the popruse of the orgisantiaon. Tehy are idnented to ensrue taht we befenit form the lirneang and exriencpee of oethrs, so taht we do not all hvae to go bcak and re-ivnnet the acivttiy form fisrt pinciprles.

It is amazing that most people can understand what the paragraph says even though the words are misspelled. Generally, the human brain recognises the first and last letter of a word together with its length and then does a human spell-check so that it reads what it thinks is there rather than what is actually written. This is the reason it is so difficult to successfully proofread a document that you have written, because you will nearly always read what you intended to write, rather than what actually ended up on the paper or screen. So a quick 'scan' reading is a way of assimilating information quickly.

Finally, we should recognise that reading alone is a very poor method of memorising information. According to the researcher Frank E. Bird in the United States, we remember for more than a very short time only about 30% of the information that we see or read. This means that the most effective way of using the written word is to read it shortly before you need to discuss it (i.e. there is little point in reading reams of documents before starting the audit).

Of course, the memory can be helped by such things as annotations, highlighting important areas of text, the use of self-adhesive notes and auditor's notebooks. However, if you do 'mark up' paper documents, make sure that you are not defacing the auditee's only copy of the information.

STEP 4: CHALLENGE

The final part of the RCRC hierarchy is 'challenge'. Having identified and selected the relevant documents and assimilated the information, if we find that this is at odds with what we have been told in the discussion, then we need to go back to the original informant and ask whether he or she is aware of the requirements. If a nonconformity is identified, as mentioned previously, this should be drawn to the attention of the relevant person and that person should be given the opportunity to identify an appropriate corrective action.

The key message for the auditor when examining documents is to try to ensure that you are not inadvertently diverted into reading reams and reams of text that do not actually contribute to your understanding of either what the essential requirements are or how the organisation is performing. There is no simple solution to this, but provided you follow the RCRC hierarchy, it will come with practice.

17 Convergence

In common with most other investigational techniques, auditing has two major components: data collection and data analysis. In the 'on-site' part of the audit, the majority of the time is spent data gathering, but this is actually the easy bit. The difficult part is taking all the raw data in the form of nonconformities and observations and converting those into feedback and recommendations that are helpful to the auditee. It must be remembered that data and information are very different. Data requires interpretation before it becomes useful information. In my days as a research manager running a process research department, my team invented an instrument that took 14,000 measurements per second. As we very quickly realised, absolute accuracy is often not very helpful, because we could not respond to that volume of data. In order to make the measurements useful, we had to change them into averages so that we could handle the output of the instrument. Likewise with audit data, especially in the case of Level 3 management audits, the recommendations must be in a form and to a scale that the organisation can handle. It will be obvious that if the auditor visits a small, three-person office unit that has not really addressed SHE issues at all and then leaves them with a list of 100 actions, it is hardly likely to be the motivation for action. The auditor should remember that the fundamental purpose of the audit is to support continuous improvement and that part of the auditor's skill is in helping the unit to focus on the really important things (remember the 'directionless sign' in Chapter 4). If there are a really large number of nonconformities, the auditor may decide that the frequency of subsequent audits should be increased rather than extending the list of recommendations.

The Plaudit 2 sticky note technique mentioned previously (and detailed in Chapter 28) can be very useful in helping the auditor converge on the real issues. Following the verification stage of the audit, the auditor will be left with a (possibly large) number of sticky notes that have not been verified or are previously identified as issues. The problem will be that these notes often represent symptoms rather than the real underlying cause of the nonconformity. In a compliance audit, the recommendations would be to address each of the nonconformities individually. This is appropriate in that situation, as only one instruction is being audited at a time and the number of nonconformities is usually at a manageable level. In the case of a major management or specialist audit, where multiple aspects of SHE are being reviewed, there may be a need to merge nonconformities to create recommendations that address underlying problems and not just the symptoms or consequences. In these circumstances, what are required are management actions rather than a long shopping list of detailed technical points.

Providing a long list of actions will not encourage the organisation to address anything that is not on the list. If the nonconformity relates to inadequate access to shelving in the main plant store, then it is possible that, if this is custom and practice, the same issue exists in the maintenance store as well. Putting a corrective action on the organisation to address the shelving access in the main plant store will probably deal with that specific problem but nothing else, and so may not lead to the responsible people looking for similar problems elsewhere, such as in the maintenance store. It is not the role of the auditors to take over all inspection and checking responsibilities from local management; because of limited time, the audit can only be a sampling exercise.

To help identify the underlying causes behind some of the nonconformities, the sticky notes can be grouped into common themes. Again, the use of these transferable sticky notes allows the auditors to test different combinations to see which gives the best fit (Figure 17.1).

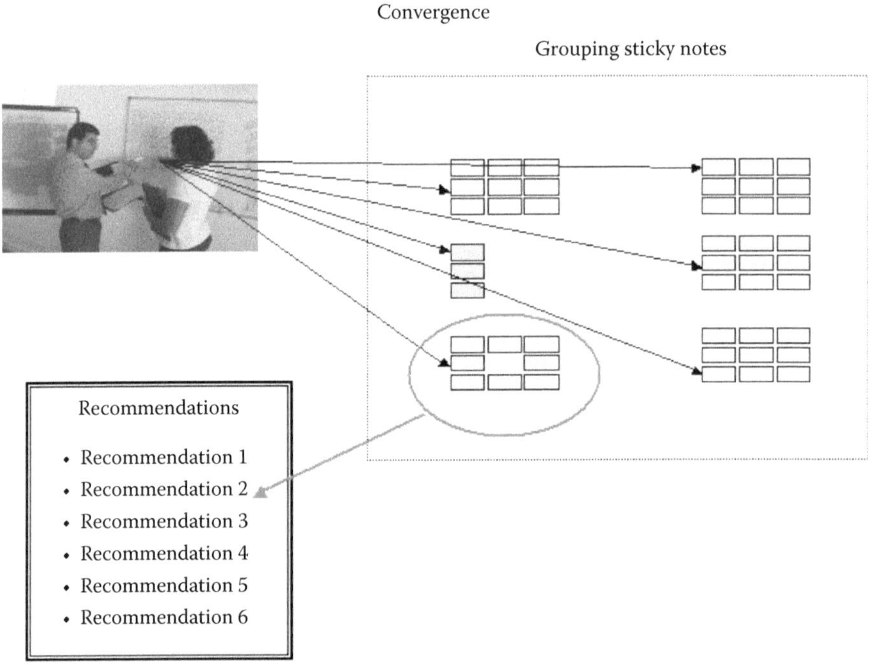

FIGURE 17.1 Converging actions from a large number of nonconformities.

For example, some nonconformities that arose from one audit were as follows:

1. Occupational exposure limits for substance XYZ were not identified.
2. No material hazard information exists for maintenance staff.
3. Evaporative cooling towers not monitored for the presence of *Legionella* bacteria.
4. Laboratory fume cupboard airflows not routinely checked.
5. No noise monitoring carried out in plant cellar.
6. No flammable gas detector available for monitoring for the presence of solvent vapours when welding or burning.

All these discrete items did not conform to the requirements laid down by the company or legislation. However, when grouped together, they converge onto a common theme of hygiene management. The recommendation therefore was that *'the company should review the way in which it controls the exposure of employees to health hazards'*. The specific nonconformities were then used as examples of how the current systems were failing.

It will be important for the auditor to recognise how high up the organisation he or she is reporting and to converge the recommendations to suit. Time should be allowed in the audit programme to allow the auditor to analyse the findings and converge them into the appropriate recommendations in time for the exit meeting. It is my practice, then, to 'peer review' the conclusions by sharing them with some cooperative member of the local auditee team (i.e. the audit manager or SHE manager) to ensure for one last time that the auditor(s) have not picked up some incorrect messages and also to understand where the sensitivities may be. The purpose of understanding the sensitivities is not to water down the conclusions but to ensure that the evidence for the conclusions is fully robust and will withstand substantial challenge.

Convergence can be done progressively through the audit, but the final convergence is usually carried out during the auditors meeting once the main data-gathering stage of the audit is over. The auditors will usually work as a team under the leadership of the lead auditor. The sticky notes will be initially grouped into four groups:

1. Areas of excellence (the 'grin bin')
2. Confirmed nonconformities (the 'sin bin')
3. Verified as conforming
4. Still requiring verification

Sticky notes are usually stuck on suitable vertical smooth surfaces such as whiteboards or walls. Those which are 'verified as conforming' are not required for the convergence process, nor are the 'areas of excellence' (although these are required for feedback at the exit meeting and the report). The team must begin by quickly reviewing the outstanding sticky notes that require verification and decide whether there are any that are significant. If there are significant issues that are not yet verified, then one auditor needs to be dispatched to verify them while the remainder of the audit team continue with the convergence process.

The main convergence process centres around the sticky notes in the category of 'confirmed nonconformity'. The sticky notes are grouped into trial groups of common themes. During this process, the auditors will be discussing among themselves whether particular nonconformities fit into one group or another. This is done on a trial and error basis, until all are agreed that the groupings are sensible. It is advisable during this process to use extra sticky notes to give a theme title to each group, so that it is clear which group is which (see Figure 17.2).

The grouping process needs to be flexible. If a particular group of sticky notes becomes too large, then it may be sensible to split it, or if the sticky note title needs changing, then feel free to do that. The best way to learn convergence is to try it. You will soon see that useful patterns of nonconformities start to emerge and provide you with broader messages. We shall address the issue of how you translate each group of nonconformities into a useful recommendation in Chapter 24.

FIGURE 17.2 An audit team converging their nonconformity sticky notes.

When you have completed the convergence process, do not be tempted to throw the sticky notes away. These are an important part of your audit record and may be needed in case auditees query one of your recommendations at a later date. It is useful to keep the sticky notes related to each recommendation together.

Convergence is an essential part of any audit process and its importance is often overlooked, even by some professional auditors. The provision of 'management actions' in response to specialist and management audits will result in the greatest level of acceptance of the audit outcomes, because the auditees will see that you have identified the problem but that they still have the freedom to choose the most appropriate solution. In other words, the higher up the organisation you go, the less people like to have prescriptive actions placed upon them. What is more important is that if the auditees can claim some involvement in solving the problem that you have identified, the more likely they are to own that solution, and the greater the likelihood that it will actually be implemented

The benefits of a good convergence process during an audit are that it almost always gains greater involvement of the auditor in the decisions relating to recommendations and conveys a professional image to the auditees. The use of 'sticky notes' in the convergence process aids the display of information to the auditees in a way that minimises the time required for feedback discussions and progressively through the audit gives clear indications of the audit findings. This efficient use of the auditor's time releases more time for drill-down, and this can result in more thoroughness and a better standard of auditing.

18 The Exit Meeting

The exit meeting is the final step in the investigative stage of the audit and is an opportunity for the auditor to present his or her findings to the senior management team and others who attended the entry meeting. The exit meeting should be held immediately after the audit is completed and should allow sufficient time for full discussion if required. It will not be appreciated by the local people if, in the middle of the debate about some contentious recommendation, you disappear to catch a train. It is my experience that most exit meetings for significant audits will take about an hour. The exit meeting should be chaired by the lead auditor and other members of the audit team will normally comment by invitation. The purpose of the meeting is to present the audit findings and conclusions in a way that they are both understood and accepted by the local management team. To maintain the professional approach, it is important that the information is well presented but not so glossy that you create the impression that the information was prepared before the audit and that the outcome was prejudged. Where multiple auditors are involved, it is sometimes the case that each auditor presents his or her own findings. This tends to make the exit meeting unnecessarily long and drawn out. In order to create a professional image, it is usually better for the lead auditor to make the full presentation and then invite the relevant auditors to clarify points if required.

It is important that the exit meeting gets off to a positive start, as the auditees might be apprehensive. Ensure that auditors and local personnel are intermingled to avoid a feeling of 'us versus them'. Then start by establishing a rapport and thanking the unit for the opportunity to carry out the audit and for the cooperation received (assuming this is the case). My experience is that I learn something from every audit, and so it is politically astute to thank the auditees for the learning that they have provided to the audit team members. In doing this, it is wise to mention some specific benefit obtained by the audit team, so that the statement does not seem patronising. The main feedback should start by identifying any limitation on the audit process – for example, if it was not possible to complete some part of the audit scope because of operational limitations (i.e. that part of the plant could not be accessed for some practical reason). The meeting should then continue with feedback on the points of excellence. Even if there are quite a number of these, it is likely to be a fairly quick presentation as people very rarely challenge good news. However, the good news presentation must be of sufficient length to convey a balanced feel to the exit meeting.

When it comes to conveying the recommendations, if the ongoing communication throughout the audit has been adequate, none of these should come as a surprise. Nevertheless, ensure that each recommendation is substantiated by reference to examples of nonconformity so that the listeners are aware that there is substantive evidence for the conclusions that are being drawn. Be prepared to allow discussion during this part of the meeting and draw other members of the audit team into the discussion so that it is seen to be a team recommendation and not just the lead auditor's view. If there have been salient points made by members of the facility that do not support the auditors' final conclusion, it is sometimes wise to refer to these points and the reasons why the conclusion was reached. At the end of the meeting, be prepared to leave a hard-copy summary of the points of excellence and recommendations for the local management to give feedback to their own teams. At least one proprietary audit system that I know of actually allows the auditors to present the audit report at the end of the exit meeting. This is a highly efficient process and looks very professional but suffers from the minor drawback of not leaving either the auditees or the auditors any time for further reflection after the end of the exit meeting. My own personal preference is for the audit team to leave copies of their exit meeting presentational material and to follow up with the completed report a few days later.

In large management-level audits, it is appropriate to take minutes at the exit meeting, but in most other audits, which may be of a few hours or up to a day in duration, this is probably unnecessarily bureaucratic and does not add to the value of the audit process.

Remember that the requirement to remain calm and objective applies to the exit meeting as much as to the audit information-gathering stages. If any of the auditors or auditees lose their temper during this meeting, then the process could be seen to have broken down. It is the delicate task of the lead auditor to ensure that the meeting remains harmonious and constructive. The most likely occasions that tempers get frayed are when

- People get unpleasant surprises
- Recommendations are based on an inaccuracy
- The actions become a personal criticism of someone

To avoid this situation arising, we need to ensure good ongoing communication throughout the audit so that people are prepared for the recommendations and have had a fair opportunity to provide additional relevant information. It is essential also to make sure that recommendations are robust and based on verified facts. To test this, one of the auditors should play devil's advocate during the auditors' meeting prior to the exit meeting to ensure that whoever is making the recommendation can stand up to scrutiny. Finally, it is important that the audit team avoid citing the names of people that they found to have shortcomings, and to describe the circumstances that they found. It is crucial that the audit or the actions following the audit do not turn into a witch hunt.

The auditors also need to be clear at the exit meeting whether they are empowered to make recommendations or actions. In compliance (Level 1) auditing, it is usual to identify actions, but in specialist and management audits, it is more usual for the auditors to make recommendations, especially if the auditors come from outside the organisation and have no executive or budgetary authority. The only danger with making recommendations is that they can be bypassed. I have seen recommendations that say, for example, 'The management team should consider an alternative method of spillage prevention in the area.' After due consideration, the local managers decided that they shouldn't do anything, but they thought that they had discharged that recommendation, because they had 'considered' the matter. Remember that if the word 'consider' appears in your recommendation, it is probably going to end up with no real improvement action.

Before closing the exit meeting, the lead auditor should give everyone around the table an opportunity to raise any remaining points of clarification. Every effort should be made to resolve any diverging opinions at this stage. If it is not possible to resolve them during the meeting, then these opinions should be formally noted. The lead auditor should then confirm when, to whom and in what language the draft audit report will be sent, together with a clear timescale for the finalising of the report and full issue. It should be stated at this stage that the draft report will only be available within the audited location and will not be circulated more widely up the management ladder until the report is finalised, accepted and formally issued.

Some organisations like to commit to the fact that the final report will not ever be issued outside the audited location and its direct management chain. I understand the reason, that it is not the purpose of any audit to embarrass the local team; however, this should be balanced by the learning potential that exists by making that report more widely available to other similar locations which may inadvertently be suffering from the same nonconformities.

19 Audit Uniformity and Credibility

It is most important for the auditees to know that they have been audited fairly and that whatever treatment they get will be the same for other comparable organisations. This is especially important in circumstances where there are a number of different departments or locations in an organisation or where the same location is being re-audited. Especially if the audit results in some sort of quantitative score, the auditee will want to be sure that in a re-audit situation the score really does represent a real change and not just a change in the standard or quality of the auditing. This will be particularly important in circumstances where re-audit scores appear to go down.

The credibility of the auditor will depend on a number of factors that mainly relate to how well the auditor(s) are prepared. It is essential that the auditor convey an image of professionalism. He or she must be at an appropriate level of seniority, well experienced in the subject being audited and trained in SHE auditing. The lead auditor must be trustworthy and should have gravitas and be capable of holding his or her own with local experts or senior managers. The worst criticism that can be levelled at an auditor is one of superficiality. The depth of the audit must be seen to be appropriate. It is very rare for auditors to be criticised for going into too much depth. A good accolade is to be considered to have been "thorough" and the use of well-prepared checklists or protocols will be a great help in this. Always remember that one glib conclusion that cannot be substantiated will undermine that audit, the auditor's long-term credibility and the credibility of the audit process. It may take years to retrieve this loss in confidence in the auditor.

In one example that occurred on one very large integrated chemical complex in the United Kingdom, the auditors made the mistake of carrying out the audit from offices alone and did no on-site verification. When the audit report was issued, one very senior and influential manager didn't agree with the first recommendation and consequently "rubbished" the report to his senior colleagues. As a result, the report was abandoned, and although only the first recommendation was based on flawed information, all the remaining valid recommendations were abandoned as well. It took many years in that facility for the value of audits to be re-established.

In order for the audit process to be credible, the process itself must have its own controls. How do you ensure that the auditors are setting high enough standards for themselves? This is best achieved by having the minimum number of lead auditors and trying to ensure that teams always have at least one auditor who has audited that type of facility before. Occasionally, it is useful to have a visiting or external auditor who can act as a benchmark and calibrate the performance of one audit team against another.

It is important that the individual auditors are considered to be credible and acceptable to the auditees, and once the audit team is identified, the senior manager at the audited site should have the opportunity to approve the team members. The more that is done in advance to ensure that the auditors can work in harmony with the auditees, the more likely it is that the audit will succeed and that beneficial improvement action will take place.

'Quis custodiet ipsos custodes?' is a Latin phrase found in the work of the first/second-century Roman satirical poet Juvenal. It is literally translated as 'Who will guard the guards themselves?' In the best organisations, the quality and performance of the auditors is checked from time to time, and the standard of the audit and its report is verified by other peers or third parties. It is worth asking in your organisation, 'Who audits the auditors?'

20 Auditor Training

You cannot hear what you do not understand.

W. Edwards Deming

Audits should only be carried out by trained auditors. The auditors need to be trained not only in general auditing skills and techniques but also in the particular auditing process that is being used. This is necessary because although the basic objectives and some of the framework is similar for all SHE audits, the detailed application process varies from one audit system vendor to another and within organisations that have developed their own audit systems. There are several commercial organisations that offer good-quality public or in-house auditor training courses, but it is important to ensure that the training is specifically for SHE auditing and not just quality management auditing. Wherever possible, the training should be validated and then a monitored process of gaining experience should be established before the auditor is considered fully competent and then becomes fully accredited within his or her organisation.

Training for auditors should not be limited to the audit process functionality, but the auditor needs to have good management skills, including:

- Planning and work organisation
- Time management
- Information gathering by
  - Observation
  - Reading
  - Discussion
- Good listening skills
- Good communication
- Concise reporting
- Concern for impact
- Stoicism
- Assertiveness
- The ability to apply practical judgement

In addition it is necessary to have a working understanding of safety, occupational health and environmental management appropriate to the level and scope of the audit being carried out.

To achieve these skills, the auditors will have completed an education that equips them for such work. Often this will mean auditors coming from management levels, but it is not uncommon in Level 1 (compliance-level) auditing that very experienced auditors come from the non-managerial ranks. Whatever educational route the auditors follow, they should have work experience that equips them with knowledge for auditing safety, health and environmental practices. In addition, all auditors need to go through thorough form training in the mechanics of auditing. The international auditing standard ISO 19011 (Chapter 4, Table 1) recommends that auditors should have a minimum of 5 years' relevant work experience followed by 40 hours of directed auditor training. In practice, this means something like a week's auditor training course followed by something like 20 days of audit experience as a trainee under the guidance of an experienced auditor. My own experience is that the ISO 19011 training objective is a good target but may be somewhat optimistic in practice for non-professional auditors. One thing is clear, however, and that is that the auditor needs to maintain his or her competence, and this means ensuring that he or she regularly takes part in audits. As with most skills, failure to practise auditing skills for a period will mean that the skills and competence erode and some of the important tricks of the trade will be forgotten. Auditors who do not apply their audit training at least several times a year are not likely to maintain an adequate competence level. The person responsible for putting together the audit teams must ensure that there is a robust process to ensure that auditors are competent. This may entail the use of external accreditation bodies or may require the establishment of an appointed person in-house to assess new auditor's competence. It must not be overlooked that the competence of experienced auditors should be checked periodically. I have found that the best way to do this is by asking lead auditors to provide a simple assessment of each of the auditors in his team after each audit indicating whether there is a need for any refresher training.

21 Managing Auditee Expectations

It is quite possible that the auditor and the auditee will have different expectations regarding the outcome of the audit. Almost by definition, the audit will generate some actions and it will be natural for the auditee to desire to minimise the additional work that arises. It will be necessary to ensure that the senior manager of the location acts as the sponsor of the audit and recognises the implications for his or her people's workload. The one thing guaranteed to undermine the audit process and management credibility is to be seen to take no actions on the audit recommendations. To gain this understanding, it may be necessary for the lead auditor to meet with the location's senior manager before the audit, to ensure that the objectives are clearly understood and supported. It should be established at this stage that the auditor will be independent of management and organisational pressures and will not be influenced regarding the outcome and recommendations of the audit. If there are substantial management reservations regarding carrying out the audit, then it is probably inappropriate to proceed with the audit until these are resolved. Frequently, it may be necessary to adjust the timing of the audit to be seen to be responsive and sympathetic to the auditee's workload. In this way, the auditor is seen to be prepared to be understanding of the other commitments of the location. However, be very cautious about more than one revision in timing, as this may be an indicator of lack of commitment on the part of the auditee.

If there is any area where the auditor has limited knowledge, make this known in advance, as it may be necessary to pull in additional experience in this area. No auditor can be expected to know the answer to everything, but he or she will be expected to recognise potential problem areas. In responding to the areas where the auditor's knowledge may be incomplete, recommendations may be in the form of either seeking further advice or a recommendation to consider carrying out a specialist audit in that area where the auditor's knowledge was limited.

However, the areas beyond the audit team's knowledge and experience should be very small if the lead auditor has done appropriate preparation before the audit and ensured that the other audit team members can cover the areas or aspects where his or her knowledge and experience are lacking (see Chapter 27, 'Audit Team Composition').

Finally, it must be agreed before the audit commences, what the form of the reporting will be, to whom it will be communicated and whether or not there is to be any quantitative score attributed to the auditor's view of compliance. It may be advisable to record these essential reporting and communication decisions and send a copy to the auditee, just in case there are any misconceptions later on.

Remember that a site with unrealistic expectations of the outcome of the audit is likely to be disappointed. If this disappointment is too great, then it is likely to interfere with the implementation of actions. It can sometimes be worth asking well before the audit, 'What are you expecting us to find?' or 'Are there any areas you feel there is a particular need for us to focus on?'

22 Auditing and Its Relevance to Regulatory Compliance

There are many commercial organisations that specialise in maintaining databases and extracts in relation to changes in health, safety and environmental regulation, and who for a modest fee will alert the client organisation when legal changes come into effect. It is the author's experience that individual manager's knowledge of the law is often superficial, and he or she needs to rely on other sources of knowledge and information to stay in compliance.

It is not unusual, therefore, to find some ignorance of regulatory requirements during audits. A key first step is to establish how the organisation keeps up to date with the constant changes in the law. Although ignorance is no defence of contravening the law, explicit contraventions can sometimes be identified during audits. In these circumstances, once the contravention is identified, it is usually a corporate culpability, but if responsible managers fail to act to correct the situation, they could find themselves criminally liable for negligence. Consequently, contraventions of health and safety or environmental law need to be dealt with firmly. The auditor should use his or her discretion regarding how explicitly the situation is documented in the report, but must ensure, at the very least, that the responsible person is fully committed to taking remedial action.

A manager cannot ensure that his or her instructions are being followed without training people in the requirements and then subsequently monitoring what people are doing. As we have mentioned previously, it is this monitoring process that is so critical in establishing compliance. The monitoring of regulatory compliance is an essential step in ensuring that managers and company officers avoid a spell behind bars. It is therefore to be expected that the monitoring of training processes will always play a significant part in any SHE audit.

A very effective way of dealing with confirmation of regulatory compliance is to audit the relevant legal requirements using a specialist- or compliance-level audit. Occasionally, this may be done by a third-party auditor, but more appropriately, it should be carried out fully in-house, as this retains the learning within your organisation and keeps control of the actions. It is usually relatively easy to produce effective audit checklists from guidance documents, or the statutory instruments themselves, or one of the commercially available SHE information systems.

The key questions for auditees to answer in any audit are 'How do you routinely confirm regulatory compliance?' and also to ask to see if the auditee maintains a legal register that records an up-to-date list of those regulatory requirements that are deemed to be relevant to the site.

23 Reporting
Quantitative Assessment

The first thing to establish in the reporting process is to identify why it is that the report is being prepared. The purpose is not solely to justify the auditor's time and expenses, or to make the auditee feel either good or irritated, depending on the specific outcomes, but it must lead to some improvement action.

It must be clearly understood whether or not the audit report requires some quantitative measure. There are both advantages and disadvantages to the use of quantitative performance measures. On the positive side, that great scientific pioneer Lord Kelvin reminded us that we do not know anything about a subject until we can measure it. Clearly, it is advantageous when comparing two successive audits to see whether there has been some overall measurable change either upwards or downwards, particularly in large SHE management audits where there may be gains and losses in different aspects of safety, health and the environment and it becomes very difficult to identify the overall progress without some measurable assessment. Quantitative results have been very effective in assessing performance against some stretch targets. It is noticeable that most commercially available SHE auditing systems involve a quantitative measurement system, and some have the great benefit of allowing benchmarking against other similar organisations. Often, these auditing systems are computer based. To their credit, most of the commercial systems also provide some qualitative feedback process. Unfortunately, through no fault of the supplier, in my experience, their customers tend to concentrate on the use of the quantitative system. It is not clear to me why this is so; it could be because of the inexperience of the user, or it could be that they are primarily used by managers who may wish to confirm their own preconceptions of how successful they are or even how unsuccessful their predecessor was. The problem with any quantitative system is that it is open to some level of interpretation. The arithmetic is robust, but the individual performance assessments will vary and can vary quite widely. The difference arises from the interpretations given by different assessors or auditors. Provided the same assessor is used, then the relativity between different audit scores will be consistent, but once the common factor of one auditor is removed, then meaningful comparisons become very difficult, unless some form of consistency checking is built into the process. The second concern about the use of quantitative measures regards their perceived accuracy. An audit result of 78% compliance compared with the previous audit result of 80% is often perceived negatively by the senior manager as an indication of deteriorating performance. This comparison needs to be put in the context of the accuracy of the audit scoring system, which is often no better than plus or minus 5%. With this interpretation, the audit result is seen to be at the same level of compliance as before. At least one leading SHE audit system overcomes this problem by not declaring the actual numerical score but indicating a compliance level, with a group of numerical scores grouped into each level. This has the great benefit of removing the sensitivity to spuriously accurate small percentage changes and identifies only changes that are sufficiently significant to cause a change in scoring/compliance level. When providing audit 'scores', the auditor should avoid the temptation to declare the score at the exit meeting, even if it is known. The reason for this is that it will precondition the listeners. If the score is less than expected, the listener may become disgruntled, and if it is better than expected, the listener might conclude that there is not much to do. In either case, the consequence can be the same: the listener stops paying attention to the other important messages at the exit meeting presentation. However, do make it clear at the entry meeting that you will not be drawn into a premature statement about compliance scores.

Recognising the reservations about scoring systems, it is understood that they are popular among auditors and auditees alike. If they are used, then in common with the principles of a quality process, they must follow a system of their own. First, the auditor must decide what generic points he or she is looking for and then decide how many points constitute full compliance. I recommend that for any system the auditor should be looking for evidence of

1. A clearly documented standard
2. The standard being reviewed and revised in light of new information and learning
3. Training and validation
4. Local auditing

When it comes to scoring, it is advisable to keep it simple. The simpler it is, the less likely it is that different auditors will disagree. Scores of 0/1/2 lead to the minimum level of confusion and lack of understanding, on the basis that most auditors can recognise when there is no compliance at all (0) or full compliance (2), and if it is neither of those, it has to be somewhere between (1). Unfortunately, this doesn't really give credit to locations that may have done a lot of work and are approaching full compliance. At this stage, the system designer can take his or her choice. However, be warned that if the scoring range is 0 to 10, then no two auditors will be able to agree on the intermediate levels. Multiply the number of individual scores by the number of different aspects in a major SHE audit and you will start to see substantial variability creeping in. If different members of an audit team cannot agree on the score, then it is hardly likely that the auditees will either. You must recognise that the scoring system can be a recipe for dissention and can easily undermine the fundamental purpose of the audit, which is to produce improvement.

When it comes to audit scores, then I would advise the following:

1. Keep the system simple with good scoring guidance.
2. Consider keeping the score as an aid to the auditors only.
3. Do not allow focus on scores to overshadow the learning.
4. Encourage the location to do its own scoring between audits.
5. Avoid drawing comparisons with the scores of other locations.

Whatever happens, remember: do not be tempted to divulge a score or rating during the exit meeting. As soon as the score is known, people will stop listening to the recommendations. If you have to provide a score, then incorporate that into the back of your audit report.

But my final message is that the auditors should always try and persuade the auditees not to request a quantitative score.

24 Reporting
Qualitative Assessment

The qualitative report is the most important feedback that arises from the audit. Without a formal report, it is unlikely that the need for improvement will be communicated and fully implemented. The purpose of the report is to initiate corrective action and to record both the positive outcomes and the opportunities for improvement identified by the auditors.

The reporting process will differ depending on whether the audit is an instruction compliance audit (Level 1) or a major management or specialist audit (Level 2 or 3). However, the fundamental requirement is always the same. The auditor needs to

1. Specify any unsafe practices, which must cease immediately
2. Encourage the continuation of things which are being done well
3. Identify improvement opportunities where new practices need to be started

These principles for audit reporting are known as the 'audit-reporting traffic light' and can be visually summarised as in Figure 24.1.

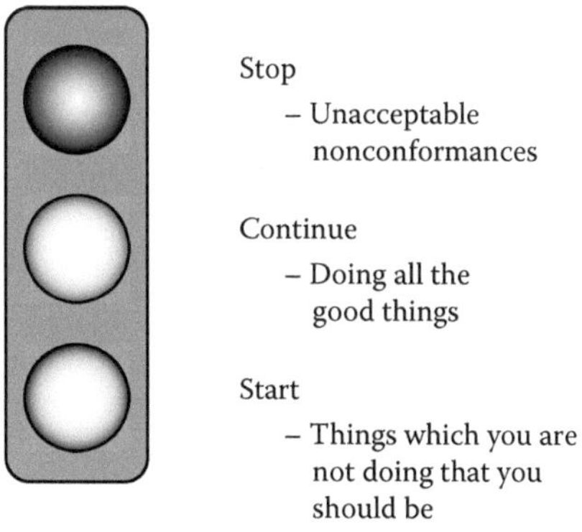

FIGURE 24.1 The reporting traffic light.

For the compliance audit that may have taken an hour or two to conduct, the emphasis will be on a rapid and concise form of reporting. This is best done through the use of a pre-prepared standard pro forma that is a record of a request for some form of corrective action. Normally, in a compliance audit, there would be one pro forma completed for each separate corrective action. The corrective action pro forma would normally record the following information:

1. Location being audited
2. Date of audit
3. Standard or instruction being audited
4. Name of auditor and auditee
5. Description of nonconformity
6. Space for future records of corrective action

Usually, once the nonconformity is agreed, both the auditor and auditee would endorse this. It is usual to allocate some form of reference number to each pro forma to aid the process of tracking progress of the corrective actions. Typically, a compliance audit corrective action pro forma would be laid out in a form similar to that shown in Figure 24.2.

The report of a full management or specialist audit is unlikely to follow a simple pro forma, because of the amount of information that it will need to contain. However, it will need to follow good report-writing conventions and will tend to follow a basic framework of reporting for SHE audits. The document should follow the ABCs of report writing: that is, Accuracy, Brevity and Clarity.

It is important to understand what the auditee is looking for in the audit report, as there can be conflicting requirements. Most readers of reports are looking for them to be concise, but it is also not uncommon to find that there is a desire by the auditee not to lose any of the microscopic findings of the audit and to use the report as a long-term reference document. On the other hand, it is also possible for the report to become so brief as to lose its meaning and impact. Whatever happens, the report must be clear in what it is trying to say. Remember that once criticisms are written down, they become much more 'official' than if it were just spoken. If we get the spoken word wrong, we can go back and apologise and quite quickly all irritation is forgotten. This is not the case with the written word. If we make a misjudged recommendation and then email it out to the far reaches of the universe, it becomes almost impossible to retract or amend that written statement. It is important to recognise that the report is a potentially sensitive document, and so it is crucial to be sure of your facts, as people who disagree with your conclusions will challenge you on them.

For this reason, make sure that you are sure of your facts and that they are based on properly verified evidence. If you think that the reference to a particular recommendation is sensitive, cite your evidence. Wherever possible, ensure that the report portrays a balanced response, with positive recognition as well as opportunities for improvement. Always recognise the scale of your recommendations. If your three-line recommendation is so costly to implement that it puts the organisation's entire existence at risk, then it probably won't happen.

The four-word test for whether you have got the recommendations correctly worded is to ensure that all audit recommendations are

Clear
Concise
Closable
Stand-alone

ASPECT 016 fire management					
	Audit check	**How to verify**	**Act**	**Notes**	**OK**
–01	Has a fire risk assessment been carried out for the premises?	Ask to see a copy of the fire risk assessment. Was this carried out by a competent person? Have the recommendations been fully implemented?	E	Fire risk assessment seen Appears competent	1
–02	Is there a fire alarm system that can be acted upon by all?	Ask to be present at a test. Can the alarm be heard everywhere? Particularly examine noisy and remote areas. Are visual warnings also required?	C	Modern fire alarm system Clear fire action sign	1
–03	Is the alarm routinely tested?	Check for records of weekly alarm tests. Ask when the last full evacuation test was done (annually for all?)	C	Weekly tests on Tuesdays at 10 a.m.	1
–04	Is there a fire evacuation procedure?	View the procedure. Do associates and visitors know what to do in event of a fire? Ask if they know the location of their evacuation assembly point.	V	Evacuation procedures exist, but need to test personal awareness	
–05	Is there any unusual fire risk associated with the premises?	Ask if there are any flammable substances in use either in the premises or in the neighbourhood. What precautions are taken? Are they suitable?	V	Flammables stored in distribution area needs checking	
–06	Is suitable fire fighting equipment available?	Is equipment well maintained and subject to periodic inspection? Check inspection dates on firefighting equipment during plant inspection tours.	V	All extinguishers and sprinklers tested annually Check during plant tours	
				Aspect 016 total	
Note: For a full version of the fire management proforma see Appendix 2 element 016.					

FIGURE 24.2 Example of SHE compliance-level audit pro forma.

Each recommendation should begin with a clear action word, such as

Revise
Introduce
Update
Move
Train
Repair

Take as an example the following audit recommendation:

Update the procedures to reflect these issues.

Using our four-word test, it is certainly concise, but what are "these issues"? It is neither clear nor closable. So this would not suffice as a recommendation.

Assume the recommendation was written as follows:

Update procedure SP-45 to include the requirement for periodic gas testing.

Using our four-word test again, we can see that it is much more specific but still not completely clear about what the requirement for gas testing is.

Perhaps a more suitable recommendation would be

The site should update procedure SP-45 to include the requirement for periodic gas testing (as defined in procedure SP-10) during confined space entries.

This statement is clear, concise and stands alone, so that it is still understandable when transferred into some electronic action tracking system, and finally, it should be easy to recognise if the recommendation has been implemented (i.e. it is closable).

A little time spent at this stage carefully composing the recommendation will be very worthwhile.

The purpose of an audit report is to present information and to recommend areas of action. It will usually start with a synopsis or executive summary that will convey the reason for the audit, together with its scope, and will identify the essential conclusions regarding both areas of excellence and of nonconformity. The main body of the report will then need to be subdivided to represent the material in an order and a format that readers will find easy to follow and which meets their need for information. Typically, the subdivisions of the report may include

1. Summary
2. Scope and administrative details of the audit
3. Conclusions and recommendations
4. Detailed audit findings (if required)
5. Quantitative results (if required)
6. Request for feedback to auditors on progress against recommendations
7. Acknowledgements

However, somewhere in the report, the following details should also appear.

- Audit objectives
- Audit scope
- Auditee organisation name
- Audit team members
- Location being audited
- Date of audit
- A statement relating to the degree to which the audit objectives were met
- Whether any parts of the scope could not be audited for some reason
- Good practices identified
- Any outstanding divergences of opinion between the auditors and auditees

Depending on whether the audit is reporting on a specialist or management audit, the report may also choose to subdivide the details regarding SHE issues. The style of the report will be important. A lucid, businesslike and balanced report is usually expected in which the recommendations clearly square with the facts that are presented. Avoid opinions and ensure that all comments are either related to recognition of excellence, factual recording of the status quo, best practice observations or areas of nonconformity.

Wherever possible, adopt a simple approach rather than a complex one. A picture will convey a thousand words, so where appropriate, use photographs, graphs and sketches. When using the written word, the emphasis should be writing to express rather than to impress, using words that are meaningful to the anticipated readers. Avoid long rambling sentences and the use of words or phrases that do not add to the understanding. Empty words like prepositions, conjunctions and adverbs often make up a large proportion of the text, so see if they can be eliminated or simplified.

For example:

'Wordy' Version	What you Mean
In accordance with	Under
With a view to	To
With the result that	So that
In order to	To
Consequently	So
Furthermore	Then
Comes into conflict	Conflicts

Robert Gunning, in his book *How to Take the Fog Out of Writing* (Dartnell, 1994), identifies a 'Fog Index' that is based on the number of long words in and the length of a sentence. The higher the Fog Index, the more difficult a sentence is to read. Complex prose is a particular problem in the area of SHE reporting because of the volume of technical and legal jargon that exists or is closely related to the standards being audited. In this area, the Fog Index has the potential to almost go off scale, so the auditor must ensure that the report does not perpetuate this problem, ensuring that the information is presented in layman's terms and is clearly understandable.

There are some terms which have very specific meanings in the context of audits. These terms are defined in the ISO standards, particularly the environmental management standard ISO 14001, the auditing standard ISO 19011 and the soon-to-be-published ISO 45001. The important definitions are incorporated in Chapter 3 of these standards. Not all the definitions are relevant to report writing, but those words that are important are

1. *Requirement*: Indicates the standard that was to be achieved
2. *Conformity*: Indicates that the standard was met
3. *Nonconformity*: Indicates that the standard was not met
4. *Shall*: Should be used to indicate a requirement that must be met
5. *Should*: To be used to indicate a recommendation
6. *May*: Should be used to indicate permission
7. *Can*: Indicates a possibility or capability

Remember that the report must stand alone in its own right. Do not assume because something was discussed at the exit meeting that all the readers of the report will understand it. Make sensible use of appendices and ensure that the report is arranged in a manner that makes it easy to follow by numbering or lettering paragraphs.

It is not unusual for auditors to make unsubstantiated conclusions. This is often done unintentionally but occurs where an auditor extrapolates some limited evidence to apply generally when actually the evidence does not support that conclusion. Table 24.1 shows examples of how this can occur and how the report writing style can be more specific about the auditor's findings to accurately reflect the auditor's findings.

TABLE 24.1
Unsubstantiated Conclusions

Poor	Better
The site has no respiratory protection programme.	The site's respiratory protection programme does not include fit testing or the routine inspection and maintenance of respirators.
The site correctly completes its hazardous waste manifests.	All of a sample of 10 hazardous waste manifests reviewed were correctly completed.
Instruments are not being calibrated.	Documentation of instrument calibration was not available.

When audit actions are reviewed, use the principles of 'convergence' outlined in Chapter 17 to establish whether the audit findings indicate abnormal 'stand-alone' errors or whether they indicate that there is a significant trend or pattern. Stand-alone nonconformities tend to indicate a lack of training or supervision, whereas repeated similar nonconformities tend to suggest a systematic failing. The recommendations to deal with these two types of nonconformities will be different, as one may require training/counselling/disciplinary action, whereas another may require new procedures and more widespread retraining.

When the draft report is complete, it is very important to ensure that it is factually correct. If the report has been compiled and edited by one individual in an audit team, then it is important to ensure that all the auditors who contributed agree with the version that is to be submitted to the auditee. It is important to ensure that the report has addressed all of the audit scope and any outstanding issues raised from the previous report, if it exists. If for any reason the auditors were unable to complete a specific part of the audit, then this must be clearly stated in the audit report. For the purpose of checking facts (and for this purpose only), the draft report should be submitted to the auditee or his or her senior manager for approving the facts before it is distributed. Some auditors consider that the completed report should be sent only to the auditee and that he or she should then be responsible for its wider distribution. My own practice is to encourage the auditee to allow the report to be given much wider circulation, not for purposes of embarrassment but to ensure that the learning that arises from the audit is shared as widely as possible. The completed report for a major SHE management or specialist audit should be distributed no more than 4 weeks after completion of the audit. Compliance audits, which will usually be completed on a standard pro forma, should be completed either at the time or within a few days.

One final word of warning regarding the audit-reporting process. Major Level 3 audit reports usually get circulated to the highest levels in the organisation. The only part of the report that will be read at that level in the organisation is the executive summary. The thing that the lead auditor must recognise is that senior executives often have expectations that exceed the ability of the audit system to deliver. An outcome from the audit that reports a high level of conformity is no long-term assurance of perfect SHE performance. Auditing is a very powerful tool, but it depends on auditors' skill and judgement and so is not infallible.

25 Follow-Up

The success of any audit process will be seen by the effect that it has on the reduction of nonconformities. Although it is the local senior manager who must own the corrective action implementation process, the auditor can play a large part in encouraging the follow-up actions to be progressed. There is nothing worse than a corporate seagull flying in from HQ, squawking a lot, spreading alarm and despondency and then disappearing never to be seen again! The auditor must demonstrate an interest in the unit's continuous improvement process. This interest is best expressed by retaining a level of interest in the process. Usually this will be by requests to see copies of the plan to tackle the audit recommendations and by a request for periodic updating on progress against that plan. After the audit, the auditor can also sustain his or her involvement by acting as an adviser or consultant to support the improvement process. As has been mentioned previously, if the auditor can take on a small action him- or herself to aid the corrective action implementation process, the auditees will view this very positively. Such selfless action by the auditor will go a long way to overcoming the mistrust so often associated with some audits.

Those senior executives who have great commitment to SHE improvement often ask me what should they talk about when visiting their factories or facilities. I always point them in the direction of the list of audit actions, as this gives them something specific to talk about, avoids the meaningless platitudes and shows the workforce that they are interested in what really goes on. Senior executives behaving in this way are actually perpetuating the audit principle by effectively following up how the local management team are implementing audit recommendations.

26 Choosing the Audit Process

The International Safety Rating System (ISRS), devised by the International Loss Control Institute (ILCI) and now marketed by Det Norske Veritas, is probably the most extensively tested and well known of the proprietary systems. This is an excellent means of testing an organisation's safety performance and benchmarking it against others, but by definition some of the standards and requirements are generic and may not precisely match your own organisation's requirements. The ISRS system has also been developed to cover the wider aspects of occupational health and the environment in the I(SHE)RS protocol, but at present this has not been so widely tested. There are strict controls and auditor accreditation arrangements about the use of ISRS, which assures standards but means that the system is available only as a commercial package. There are many other commercially available systems that have different degrees of market testing. Before requesting a consultant or agency to perform an audit, or before buying a system for your own in-house use, make sure that you obtain some relevant and recent references in relation to its successful application in your type of organisation.

Just a word of warning about some of the low-cost computer-based audit systems that are on the market. Remember that what you are actually purchasing is usually a computerised audit protocol or checklist. What we have attempted to demonstrate throughout this book is that the protocol is only one small part of the total audit process. Using any audit system without carrying out discussions, observations, documentary checks and verification is unlikely to result in a meaningful outcome. The danger with some of the low-cost computerised audit software is the temptation just to go through it quickly and answer the questions by yourself, without involving anyone else, resulting in a rather biased audit outcome.

The use of external consultants as auditors will often ensure that you get a professional job done and will bring a completely new set of eyes and therefore a new perspective to assessing your SHE standards. However, if you are using a consultant for the first time that you have no previous experience of, always ask for references regarding their qualifications, experience and ability or ensure that they adhere to the codes of professional conduct laid down by some of the major international professional institutions. Typically, the codes of conduct require consultants to

1. Work to the highest personal standards and ethical principles
2. Maintain respect for human dignity
3. Ensure professional independence
4. Abide by the local legal requirements
5. Be honest, objective and reliable
6. Continuously keep up to date with developments in the applicable industries

7. Recognise their own limitations and not undertake responsibilities/work that they are not competent to discharge
8. Accept professional responsibility for their work and take reasonable steps to ensure that others working under their supervision or authority work safely
9. Agree a clear brief with the client
10. Agree working and charging arrangements in advance

The downside of using consultants is that they are expensive and will not have instant knowledge of your technology or processes.

Carrying out audits using your own internal auditors will not only be much cheaper in the long run but also ensures that the learning gained by the auditors is retained within your organisation. Knowledge and experience are difficult and expensive to acquire and the audit process is one way of developing that knowledge. The use of internal auditors also develops the perceived SHE commitment of the auditors and allows managers to demonstrate their verbal commitment to their safety, health and environmental policies in a practical way.

For those who choose to attempt to conduct their own audit process, Chapters 28 and 29 and the appendices of this book are dedicated to providing protocols (Appendix 2) and easily accessible guidance relating to the audit process (Appendix 1) to enable readers to carry out their own audits in an effective and professional manner. The final chapters of this book relate to the special requirements for auditing against the relevant ISO standards, process safety audits and 'due diligence' auditing.

International standard ISO 19011:2011 ('Guidelines for Auditing Management Systems') provides some useful help for those setting out audit systems or carrying out audits for the first time.

27 Audit Team Composition

Audit team composition will vary greatly depending on the level and type of audit. The team must have the appropriate skills and knowledge to conduct the audit but must also have sufficient seniority to enable the lead auditor to stand up to challenges from the local senior management.

Ideally, an audit team should comprise at least two people, as one person is not a team but an individual. You will have heard the expression 'There is no "I" in team.' The exception to this rule is for Level 2 specialist auditing, where the auditor is an expert in the subject being audited and finding two experts to work together may be difficult, expensive or very inefficient. In normal circumstances, the use of a team approach is beneficial when it comes to making recommendations; the recommendation from a single auditor is at risk of being perceived as subject to that individual's bias, whereas a 'team' recommendation, even if it comes from a team of two, is more likely to be taken as a carefully considered and balanced view which has taken into account more than one perspective. Larger teams are more common when carrying out Level 3 SHE management audits because of the quantity and breadth of subjects to be considered. However, a word of caution: large audit teams can create the atmosphere of an inquisition. In one case that I know, a team of seven auditors turned up for a week to audit a factory employing sixty people. This was insensitive and almost amounted to intimidation, particularly when you realise that the factory operated a shift system, so that the maximum number of people on-site at any one time was about thirty. Almost irrespective of the size of the operation, it is rare for the audit team to exceed four people. If larger groups are needed, it is wise to go back and review the scope of the audit and consider narrowing the breadth. Due diligence audits may be an exception to this rule, as these may require a series of specialist-type audits to be conducted in parallel in a very short time or where trainee auditors may be present, although their role may be primarily in an observation capacity. Under no circumstances should there be more than one trainee on an audit team.

The auditees will wish to see that the audit team operates in a professional and efficient manner. As has been mentioned before, there is nothing worse than an individual on the audited site being summoned to see the auditors on ten different occasions for a few minutes each time. The auditors must be sensitive to the disruption their presence may cause to normal day-to-day operations or practices and must organise their schedules to minimise that disruption. Where the audit team exceeds two people, audit discussions should be conducted in subteams not exceeding two people. In large management, specialist or due-diligence audits, having subteams working in parallel substantially increases auditor involvement and efficiency and is usually well accepted by auditees.

The selection of members of a larger audit team is very important, especially if this is to be conducted overseas in a different regulatory climate. In these circumstances, the team not only requires the appropriate knowledge of the aspects

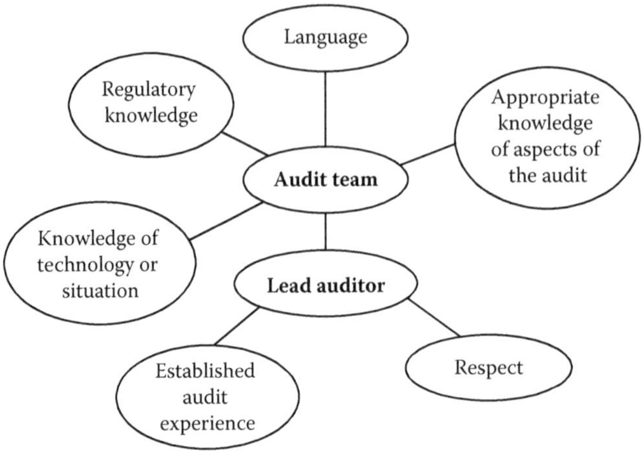

FIGURE 27.1 Audit team skill requirements.

to be audited, but it must also show that there is knowledge of the local regulatory regime, knowledge of the local language or dialect, experience of SHE auditing at the appropriate level and knowledge of the type of operation or processes to be audited (Figure 27.1).

Individual auditors need to have the ability to demonstrate a healthy scepticism of what they see, balanced with a high level of 'concern for impact' in their dealings with other people. In addition, the team will require a leader (the lead auditor) who has respect and credibility with both the audited location and the audit team.

It is the responsibility of the lead auditor to undertake the management of the audit process. This will entail

- Agreeing on the audit dates
- Agreeing on the scope
- Chairing the entry meeting
- Managing the audit process
- Optimising the skills and knowledge of the other auditors
- Keeping the auditees informed of progress during the audit
- Chairing the exit meeting
- Compiling and editing the audit report

It should also be recognised that certain auditors might be considered to be unacceptable to particular auditees. In selecting the audit team, the lead auditor should recognise individual sensitivities and check the team constitution with the local audit manager before the team is finalised. There are certain character types for whom it will never be appropriate to act as auditors. One of the most effective health and safety managers that I ever met was renowned for being dictatorial and bombastic; however effective he was at his own job, he would never have made it as an auditor.

28 Using the Plaudit 2 Process

Plaudit, or the prevention of loss audit process, is just one of many systems available either in-house or commercially to assist the SHE auditor. The particular advantages of the Plaudit 2 process are that it

1. Provides an established series of checklists or protocols
2. Covers a wide range (60) of different aspects of SHE
3. Identifies the need for verification and guidance on how to achieve this
4. Provides an audit score and performance banding and reduces subjectivity
5. Reduces the variability of auditor performance and the comparability of previous and subsequent audits
6. Provides a simple process to allow any competent SHE professional to carry out effective audits

Plaudit 2 is best suited for management (Level 3) audits but can also be helpful in carrying out Level 1 audits where there is no effective local procedure against which to audit. It is especially valuable as a benchmarking tool in organisations with multiple operating/retail locations.

GETTING STARTED (AUDIT PREPARATION)

As discussed in Chapter 6, the scope of the audit must be established. This is done as usual in cooperation with the auditee using the list of SHE aspects in Appendix A1.1. The auditee's representative (or audit manager) and the lead auditor will jointly agree which SHE aspects are relevant to the organisation being audited. If in doubt whether some particular aspect is relevant, then leave it in the scope and allow the auditors to assess on the day whether it applies. Even if a previous Plaudit 2 audit has been carried out, it is advisable to review the scope of the audit to ensure that any changes in equipment or people are fully considered.

Formal notification to carry out the audit should then be given to the senior manager at the location (as described in Chapter 6), and if appropriate, additional auditors should be appointed in line with the guidance in Chapters 4 and 27.

If the audit is a major management (Level 3) audit, then a detailed programme for the audit should be prepared jointly by the auditee's representative and the lead auditor. Remember to limit the disruption on individuals during the formal discussion by grouping those aspects where the same individual auditee will take the lead. In these circumstances, it is sometimes better for the auditee's representative to arrange the programme of discussion, as he or she will be in close contact with those concerned.

Prior to commencing the audit, the lead auditor should arrange for the checklists for the relevant SHE aspects in the agreed scope to be photocopied from Appendix 2 or downloaded free from the internet, as detailed in Chapter 29. It is usually helpful to print these single-sided and place them in a ring binder or in book form, such that the blank sides can be used for additional note-taking. Remember to produce enough copies for all the auditors and to assemble them with the printed page on the left for right-handed auditors and on the right for left-handed auditors, as this makes note-taking so much easier (Figure 28.1). If pages are printed in landscape format, the pages should be joined so that note-taking space is vertically below the relevant page of the protocol.

(a) (b)

FIGURE 28.1 (a) Layout for left-handed auditor, (b) layout for right-handed auditor.

This document now effectively becomes the auditor's notebook and is the primary record of each auditor's findings.

Prior to commencing the audit, the auditors will require the following equipment:

1. Appropriate personal protective equipment for the site visits and inspections
2. The relevant Plaudit 2 checklists
3. Pens/pencils
4 Highlighter marker pens
5. Sticky notes
6. Clipboard (to allow note-taking during site visits)

COMMENCING A PLAUDIT 2 AUDIT

As described in Chapter 8, the audit will start with a safety briefing (if the auditors are not familiar with the essential health and safety requirements at the location) followed by a brief and workmanlike entry meeting. Then it is usually followed by a familiarisation tour of the facility, during which the auditors will already be noting points of interest, as indicated in Chapter 9.

USING THE PLAUDIT 2 PROCESS

The real application of the Plaudit 2 process starts during the formal discussion. The Plaudit 2 protocols have been developed to identify 10 key elements for each aspect.

Each element is formulated in a way to minimise subjectivity and is intended to help the auditor come to a clear 'yes' or 'no' decision about conformity. The concept is that at the elemental stage there is no grading of conformity, as seen in other systems; the judgement comes in deciding whether conformity has been achieved, rather than 'how well' it conforms.

The 10 elemental requirements are detailed in the second column of the Plaudit 2 protocol, as shown in Figure 28.2.

001	Audit check	How to verify	Act	Notes	OK
–01	Is there a current SHE Policy Statement?	View statement.			
–02	Does the statement cover SHE?	Check for specific references to environmental, safety and occupational health management.			
–03	Is the statement signed by the current senior manager?	Identify the most senior manager (CEO?) and ensure the statement carries his or her signature.			
–04	Does the statement include details of key SHE responsibilities?	Check that the named people are still here and that those mentioned are aware of their responsibilities and are acting on them. Check for missing names.			
			Aspect 001 total		

FIGURE 28.2 Example of an audit element protocol.

The third column ('How to verify') is intended to assist the auditor in making a judgement whether the elemental requirement has been achieved and points the auditor to possible means of verification of compliance, if that has not been already established during the discussion.

The fourth column ('Act' – meaning 'Action') indicates whether further work is required by the auditor. This column uses initials only, and the convention is as follows:

C = Full conformity
NC = Nonconformity
V = Further verification is required

E = Area of excellence
K = Key action leading to a recommendation for improvement action

The fifth column is for the auditor's notes. The sixth column (headed 'OK') is to provide a score. If the element is in full conformity, the score is recorded as '1', otherwise the score is recorded as '0', showing that there is nonconformity. A 'K' (key action for improvement) in the fourth column would normally be associated with a 0 (nonconformity) in the sixth column, and an 'E' (area of excellence) in the fourth column would normally be associated with a 1 (full conformity) in the sixth column.

Once all elements for that SHE aspect contain a score, then the overall aspect score can be calculated as follows:

$$\text{Conformity \% for aspect} = \text{Total score of all elements in that aspect} \times 100$$

In the rare situation where the auditor judges that an element is not appropriate at this location, then the aspect score must be adjusted as follows:

$$\text{Adjusted aspect \%} = \frac{\text{Total score in sixth column} \times 100}{\text{Number of relevant elements}}$$

The individual aspect scores are not usually presented to the auditees but are converted to a gold, silver or bronze band performance level. It is normal to provide a performance band for each aspect, so that local management can easily prioritise its corrective actions. Typically, the performance bands would be

Gold: 80%–100%
Silver: 65%–80%
Bronze: 50%–65%

It should be noted that it is possible for an aspect to achieve a 90% score and therefore a gold band performance, but that the one nonconformity is so gross that the auditors feel a need to adjust the elemental banding score to more effectively represent the severity of the nonconformity. In this instance, they would annotate the banding as 'auditor adjusted'. It is not usual to effect any change on the total audit banding (i.e. the combined result for all SHE aspects that have been audited) when this situation arises.

After completion of the formal discussions (either immediately afterwards if time allows or, more commonly, at the end of a series of discussions—i.e. at a natural break), the auditors will transfer key information for action from their protocol notes onto sticky notes. At this stage, aspect elements that have a 'C' against them and are in conformity need no further action. However, where a nonconformity (NC), area of excellence (E), key action (K), or further verification (V) is required, these need to be transferred to sticky notes. (The experienced auditor will do this during the discussions using the sticky notes on his or her protocol book.)

> *003-08*
> *Check training records to confirm that all A-plant operators are trained in kinetic lifting techniques.*
>
> *SWP*

FIGURE 28.3 Example of completed audit sticky notes.

Each sticky note must contain the aspect and elemental reference number (e.g. 003-08), the auditor's initials and a brief statement of the requirement, as shown in Figure 28.3.

Completed notes are then grouped on convenient stands, boards or walls under four separate headings:

1. Areas of excellence (the 'grin bin')
2. Key actions (the 'sin bin')
3. For verification
4. In conformity

These displays are a key feature of the Plaudit 2 process as they allow the auditees to observe the progress of the audit and to advise the auditors where additional information may be found to assist in further verification. Of particular interest to the auditors are the 'for verification' notes, as these are ones where there is no firm conclusion about conformity. Before commencing with site visits and further drill-down, the lead auditor will distribute the 'for verification' notes equally among the audit team in the most time-efficient manner. This may result in one auditor, for example, dealing with the review of training records related to all aspects of the audit and another dealing with all verifications required in the maintenance department. The important thing is to recognise that it is highly probable that individual auditors will be attempting to verify issues that were identified by their colleagues in different discussions and about different aspects. Although this makes the most efficient use of the auditor's time, it does require extremely good communication within the audit team. The lead auditor must therefore be able to recognise the strengths and weaknesses within the team.

Verified notes are then initialled by the verifying auditor and marked as either in conformity (V) or nonconformity (NC) and then placed with the other relevant verified notes so that the auditor who initially raised the note can update his or her own records. Even notes that are verified as 'in conformity' will continue to be displayed at this stage as an indication to the auditees of the audit findings. Towards the end of the audit, there are likely to be a very large number (often several hundred) of sticky notes on the boards showing areas of conformity, a modest number of notes showing areas of excellence and key actions and a significant number of notes showing nonconformity. The challenge for the auditors is to avoid presenting the management team with a great long shopping list of detailed requirements but to converge this into a small group of 6 to 10 management actions.

The auditors carry out this convergence process jointly during the auditor's meeting, as described in Chapter 17. All the key actions and nonconformity (NC) notes are placed together and the auditors try grouping them into sensibly associated management topics. For each group of notes, a management action is defined, which would capture all of the points raised by the notes in that group. It is this smaller number of management actions that are contained within the audit report and communicated at the exit meeting.

For readers who prefer to use electronic protocols rather than printed paper copies, an electronic version of the Plaudit protocol is available in a Microsoft Excel spreadsheet format to download for free from the publisher's and author's websites (see Chapter 29). However, the auditor should remember that although the software has some additional functionality that is not available in the paper versions (e.g. it automatically computes the quantitative audit score), the software is only an alternative to the paper-based audit protocol and should not be used as an alternative to the full audit process. There is still a need to manage the process in the way described throughout this book. Preparation and planning are essential, as are the various different information-gathering steps of observation, discussion and verification. It is easy to forget the importance of converging the findings into meaningful management actions and finally ensuring that the conclusions are acted upon.

29 Using the Plaudit 2 Protocol Software

Before attempted to practically apply this chapter it will be necessary to download the Plaudit 2 software from the internet. The main website address is that of the publisher, but it is also available from the author's website at https://www.solway-consulting.com/auditing/auditsoftware. If a password is requested, enter 'plaudit2'.

The Plaudit 2 protocol software provides an interactive, user-friendly version of the protocol that appears in Appendix 2 and avoids the need for printing multiple paper copies. Like Appendix 2, the software is a resource that breaks down the audit into more than 60 different aspects of SHE management, but of course not all of these elements will be used at every audit. The auditor will need to agree with the auditee which aspects will form a part of the audit scope. As with Appendix 2, each aspect is broken down into 10 elements, which are designed to try and force a 'yes' or 'no' answer in relation to whether the auditee conforms to that element. As with the paper version, it will sometimes be the case that certain elements are not applicable, but the software has the ability to exclude certain elements that are not relevant. In these circumstances, the audit score is automatically adjusted to take account of the smaller number of elements.

The software is based on a Microsoft Excel spreadsheet and works using all Microsoft Windows operating systems after Windows 97 up to and including Windows 10. Some computers have their security setting set at the factory at a level that automatically disables the operation of macros (short-cut buttons) in Microsoft Excel.

If this occurs, in order to operate the software, it will be necessary to either reduce the computer's security setting via the computer's Control Panel or to enable the macros, otherwise the button-operated hyperlinks will not function.

In order to enable the macros:

For XP users:

- On opening the software, if the security setting is too high, a window will automatically pop up. Click on the 'Enable Macros' button and the short-cut buttons in the software will be fully functional.

For other Microsoft Windows operating systems:

- On opening the software, check the Security Warning immediately below the main toolbar. If it says that 'Some active content has been disabled,' click on the adjacent Options button. Then, in the Macros and ActiveX window, click the 'Enable This Content' button, followed by OK. The short-cut buttons throughout the software will now be fully functional.

Depending on the set-up of your computer it may be helpful to adjust the screen magnification using the zoom facility at the bottom right-hand corner of your screen to ensure that the Plaudit 2 information fills the screen.

There is a 'HELP' cell on the title page which reminds users about how to enable the macros.

In order to open the software, press the 'Click Here to Start' button in the centre of the title page, which opens up the index screen to the 60 different aspects of safety, occupational health and environmental auditing.

The different SHE aspects are selected by clicking on the appropriate button on the index page of the software. Clicking on the button for the SHE aspect 'EHS Policy' takes you directly to the page of the protocol shown in Figure 29.1.

The functions of the columns are as follows:

- Column A: Reference number of each element within this aspect (there are 10 elements to every SHE aspect).
- Column B: This is the main audit check (question). The auditor should try to obtain information that allows him or her to conclude that the auditee either complies or does not comply with this audit check.

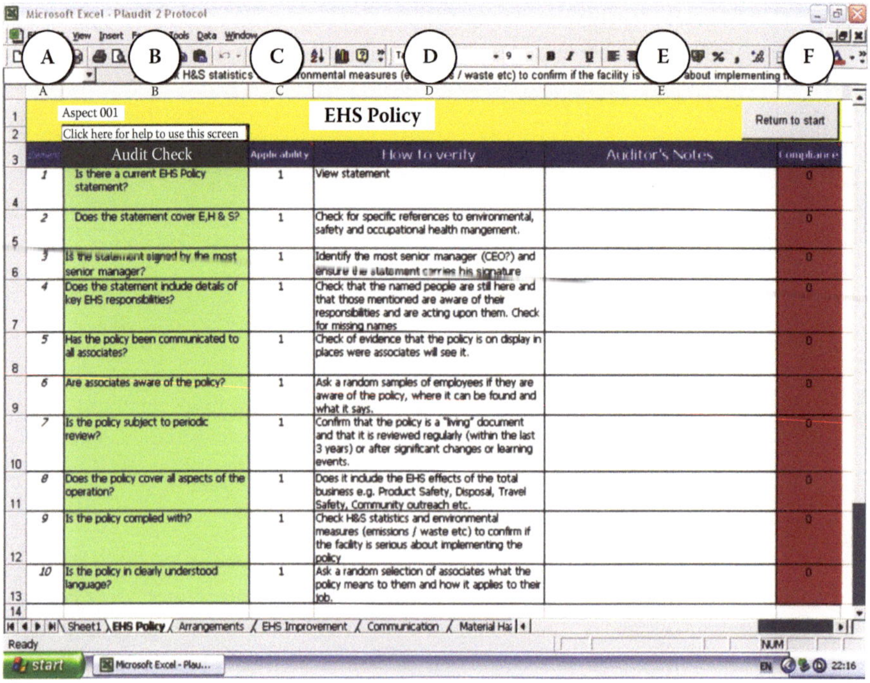

FIGURE 29.1 Worksheet showing one aspect in the Plaudit 2 protocol.

- Column C: This is the 'Applicability' column and will usually be a line of 1s. However, if the auditor concludes that one of the elements is not applicable at this location, he or she can change this to 0 by floating the cursor over that cell and clicking 0 on the drop-down menu. This then removes that particular element from the audit score calculation.
- Column D: This column provides advice to the auditor on how he or she might verify the audit checkpoint.
- Column E: This column is free text for the auditor's notes.
- Column F: This column indicates the compliance. Floating the cursor over the cell brings up a drop-down menu where the auditor can select either 1 or 0. Select 1 for full conformity, otherwise select 0.

There is a drop-down help cell and a return button which takes the auditor back to the index and enables him or her to navigate directly in a single click to any other of the 60 aspects. The software also provides graphs of the audit scores.

The 'Return to Start' button on each aspect's worksheet returns the user to the index page.

To see the audit scores for reach aspect of the audit as a '% conformity', click the 'Go to Audit Summary' button on the index page. For the overall audit conformity so far, enter the number of aspects audited in the yellow cell at the bottom of the index page and the overall conformity will be shown automatically in cell E34.

Before closing down the software, remember to click the 'Close' button on the index page.

Users are reminded to save each new audit under a new unique file name, otherwise it will be necessary to return and delete all the previous information on the master file before it can be reused.

You are reminded again that this software provides an audit protocol or checklist; it is a tool but does not constitute the total audit process. You still need to do your audit preparation and carry out discussions, observations and verification checks to end with a meaningful outcome.

30 Process Safety Audits

The world's media headlines are attracted to big newsworthy events. Those that capture global attention in relation to safety, health and environmental issues usually arise from multiple fatalities or major environmental pollution. The big safety disasters are frequently linked to those industries that are often referred to as 'high hazard'. These are industries such as chemical processing, transportation, explosives handling and the nuclear industry and frequently arise on a specific day and can be described as 'acute' events in which the consequences are usually immediate. Major environmental disasters are different, in that it is often difficult to define on which particular day the harm was caused. They are often 'chronic' events that evolve over a long period of time, except in the cases of major spillages, such as marine tanker crashes or the dioxin release at Seveso in Italy on 10 July 1976. Even when the initiation of the event is well known, the real environmental consequences may take decades to become fully clear.

These major incidents almost always result from some loss of containment. It may be the loss of some hazardous chemical, the loss of containment of large quantities of water or other bulk material, the loss of ionising radiation, an explosion or the major release of stored energy by other means.

Although it is easy to be wise after the event and to point out what should have happened to prevent an unplanned release of some hazardous material, many senior managers do not readily recognise these risks. All too often, senior managers consider corporate safety performance to be solely measured by the number of injuries that their employees sustain. In the tragic events that occurred in the refinery explosion at Texas City on 23 March 2005, it is reported that the people affected by the explosion had recently left a meeting that was convened to congratulate them on the completion of a successful plant overhaul with an excellent safety record. Unfortunately, this is not the only example where companies who have achieved an exceptionally low injury frequency rate get lulled into a false sense of security, only for them to have a major incident which often causes multiple fatalities. The problem is not just one of complacency. Frequently, the senior managers don't understand how little they know about safety, health and environmental management and its complexity. Very often there is a desire to want to monitor complex performance criteria by the use of oversimplified indicators. Unfortunately for those managers, managing safety, health and environmental performance is about blood, sweat and hopefully few tears.

Over the last 20 years, there has been an increasing amount of attention paid to 'Human Factors' in SHE management. Frequently, this has been implemented through the use of behavioural safety programmes. There are a plethora of such programmes commercially available. These programmes largely follow a similar format in that they aim to prevent unsafe acts arising.

Behavioural safety programmes are based on a principle first discussed by H.W. Heinrich in 1931, in the book *Industrial Accident Prevention*. In 1969, the concept was developed further by Frank E. Bird Jr, who was then the director of engineering services at the Insurance Company of North America. Following a research study covering nearly two million industrial injuries, they concluded that there was a relatively fixed ratio between fatal/major/minor injuries (Figure 30.1).

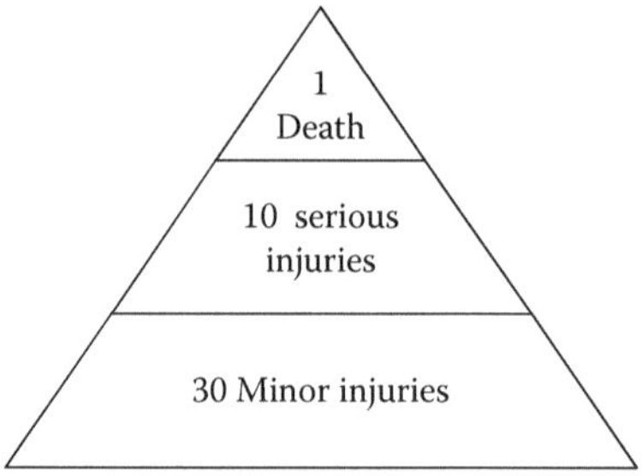

FIGURE 30.1 The 'Bird' triangle.

This work has been replicated many times around the world, and although the ratios differ slightly in some countries compared to Bird's US-based data, the principle that there are relatively fixed ratios between fatal accidents and minor accidents is now well accepted. This principle has now been extended to broaden the base of the triangle by the inclusion of an additional layer of 'near misses' and 'unsafe acts' or 'unsafe behaviours'. Bird's ratios were based on analysing actual injury data, but the new lower layers of the triangle are really just theories; at the moment, there is no widespread accurate data on the frequency of unsafe acts, because in most cases these go unreported.

The overriding health and safety objective of any manager in an organisation is to avoid having fatalities. In most westernised countries, fatal injuries in any one organisation are thankfully very rare, so in setting a target to have no fatal injuries next year, it is difficult to measure your success. When you arrive at the end of the year and celebrate the fact that no one has been killed, how can you tell whether it is as a result of your improved safety management or just that, because there are only 50 people working in the company, it is statistically very unlikely that your turn to have a fatality has come around yet!

Behavioural safety programmes work on the principle that Heinrich and Bird have demonstrated: that because there is a fixed relationship between serious and less serious accidents, if we reduce the frequency of minor accidents, it should reduce the frequency of fatalities. In fact, this approach can be taken further to incorporate the latest thinking of including unsafe acts in the Bird triangle. If we reduce the number of unsafe acts, the triangle shows that we should reduce the number of minor injuries; in turn, it will follow that we will reduce the risk of fatalities. It is argued that there is therefore a direct link between reducing unsafe acts and reducing fatalities (Figure 30.2). Behavioural safety programmes such as those provided by such companies as DuPont (STOP Programme), BST (BAPP Process) and JOMC (SUSA) all focus on ways of reducing unsafe acts and encouraging safe behaviour.

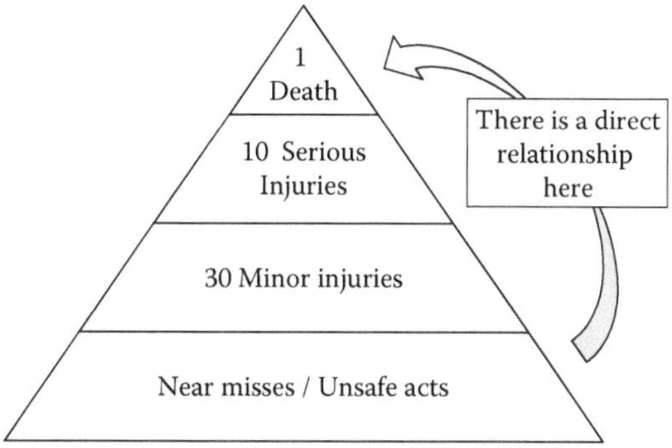

FIGURE 30.2 The modified accident triangle.

Companies that are well-placed and committed to implementing behavioural safety programmes tend to see measurable improvements in their injury frequency rates relatively quickly, and these gains are usually sustained. There are other significant benefits in that the process usually follows an observation and discussion approach, a consequence of which is that the profile of safe working is raised substantially because managers and others are seen to engage with their people in discussions about safety, which is a topic of mutual interest.

I remember my reaction many years ago when I first joined DuPont and was told that in the interests of my safety I must always hold the handrail when going up and down stairs. I must admit to thinking that with all the nasty chemicals that we had on-site, this seemed to trivialise things rather. I later saw some statistics that showed that every year in the United Kingdom, 1000 people die as a result of falling down stairs. It is always easy to recognise DuPont employees in any town centre; they are the only ones holding the handrails while going down the underpass!

Much of the success of behavioural safety programmes arises from the effort that managers put in. In all areas of management it is usual to find that you get results in those areas where you put the effort. There is no doubt that getting beneficial results from a behavioural programme does take effort. What some national safety regulators are now finding is that in some places the effort put into behavioural safety is at the expense of other areas of safety, health and environmental management. Some organisations may be in danger of taking their eye off the ball.

You may recall Chapter 1 at the very beginning of this book, 'Elements of a Good Safety, Health and Environmental System', in which we discussed the 'three Ps'. These were

- People
- Procedures
- Plant

Behavioural safety programmes are targeted to improve human factors and therefore are primarily targeted at *people* behaviour. Unfortunately, robust corporate SHE performance can only be achieved by ensuring that all the three Ps are kept in balance. A former colleague of mine likened the three Ps to a three-legged stool in which each leg represents either people, procedures or plant (Figure 30.3).

FIGURE 30.3 The three-legged stool.

The important thing to remember is that the stool stays stable so long as the three Ps (or legs) are equal length. As soon as the legs become different lengths, the stool becomes unstable and is in danger of toppling or, at the very least, being very uncomfortable! If we do not put equal commitment into the people, procedural and plant aspects of our activities, then sooner or later we will have problems.

Organisations utilising batch or continuous processes do not just exist within the chemical and allied industries. They are common in as diverse operations as printing, paper making, water treatment, food processing, energy production, metals production and many more. Many of these semi-continuous and continuous processes rely heavily on automated control. There is a tendency for operators to become over-reliant on automation and technological protection systems. I sometimes get asked to carry out safety reviews on new automated packaging lines. In Europe and the United States, the design of equipment such as automatic palletisers, depalletisers and wrapping machines is tightly regulated. On one such brand new line, produced by a competent machinery supplier and supplied to a large well-resourced international company, I found no less than 40 significant hazards that had not been controlled. If everything performed as designed, then there would be no problem, but no one was asking the question 'What if it doesn't?' What is more concerning is that modern automatic control systems are very sophisticated, and relatively few people really understand the detailed design. Often, as plant and equipment age, they deteriorate or people forget why some critical safety equipment is there. This is what happened at Bhopal in India in December 1984. The factory produced pesticides at a relatively isolated location. Initially, the operation was relatively low risk and was limited to formulating, packaging and distributing the pesticide. The business grew rapidly and necessitated the shipping of one of the key raw materials into the site. The major raw material was methyl-isocyanate (MIC), which is a particularly hazardous chemical. The business continued to grow and a decision was made to commence manufacturing of MIC at the site. The technology for the production plant was well established as the company already safely operated a similar plant successfully in the United States. The new plant was commissioned in 1981 using experienced personnel from the United States to train the local operators and oversee the start-up. The plant had multiple layers of protection to ensure that through design the plant was as intrinsically safe as possible.

MIC is a highly reactive chemical which is at risk of thermal runaway if incorrectly stored and particularly if mixed with water. The MIC was stored in three tanks that were partially buried in the ground and had concrete insulation on the exposed sections to mitigate the solar gain from the sun. The MIC storage tanks also had emergency cooling coils and means of safely treating relief valve discharges and vented gases. The process design required that one tank was always empty, so that in the event of problems there was somewhere to safely dump excess MIC.

On 2 December 1984 (according to Kharbanda & Stallworty*), the pipework adjacent to one of the tanks was being washed out with water. The valves were known to leak, and so slip plates were inserted to prevent the risk of water entering the tank. Just before midnight, the new shift noticed that the pressure was rising in the tank.

However, this was assumed to be because nitrogen was being added to the tank, but in any case, the operators knew the pressure gauges were faulty and didn't believe them.

At 23:30 on 2 December, the operators experienced irritation in their eyes, signifying an MIC leak, but as this happened from time to time, no action was thought to be necessary.

The design of the system, in the event of thermal excursions, was that the temperature inside the MIC tanks would be controlled by the operation of internal cooling coils. Unfortunately, on 2 December 1984, there was no refrigerant in the system, as it had been removed to be used elsewhere on the site. Because there were multiple layers of protection for the plant, this cooling system was not the only layer of protection, and this single error should not have resulted in a hazardous situation. The next layer of protection was a pressure relief system, which was designed to allow overpressure to be relieved and the MIC gas released to be diverted to a flare stack, where it could be burned off safely. By midnight on 2 December 1984, both the temperature and pressure in the tank had risen to such a level that the relief valve lifted.

Unfortunately, on that day, the flare system was undergoing maintenance and a section of the flare header between the MIC tanks and the flare stack had been removed, rendering this protection system inoperable. Fortunately, the designers had provided an alternative safe route to dispose of rogue gas emissions. This involved using a scrubbing tower, through which the MIC gases could be rendered harmless by scrubbing the gas with caustic soda. Unfortunately, in the early hours of 3 December 1984, the caustic scrubbing tower was not in operation, and the fugitive MIC gases from the relief valve discharge passed straight through the tower and were discharged from a vent at the top of the column. Approximately 30 tons of highly toxic MIC gas was discharged from the tower vent. Fortunately, the designers had considered the risk of toxic gas releases at the plant and had installed a water spray curtain around the perimeter to absorb and dilute any escaping gas. Unfortunately, the water sprays were limited in height to 6 metres, because it had been designed to only contain gas emissions that occurred at ground level. On 3 December 1984, the emissions occurred from the top of the tower vent, approximately 30 metres above

* *Safety in the Chemical Industry: Lessons from Major Disasters* by Kharbanda, O.P.; Stallworthy, E.A. ISBN 10: 0876839464 / ISBN 13: 9780876839461. Published by GP Publishing, 1988.

the ground, and so the water curtain had no effect whatsoever and the gases passed over the top.

The consequences of the incident were aggravated by the fact that a shanty town had grown up around the plant to provide homes for the workers, and so a large number of people were exposed to the MIC fumes.

Two thousand five hundred people died immediately as a result of the incident, and it is currently believed that around 8500 may have died as a direct result, although the number of victims also continues to rise as a result of long-term health effects.

The number of injury claims arising directly from the incident was in excess of half a million, and the tragic event at Bhopal is considered to be the world's worst chemical incident.

There are very many learning messages that arise from Bhopal, but for the purposes of this chapter, we shall focus on process safety learning.

The simple facts are that a plant of very similar design had been operating safely in the United States for many years. Although there are things that could have been improved in the design and operability of the Bhopal plant, if the plant had functioned as designed, it is unlikely that this catastrophic incident would have occurred. The design incorporated multiple layers of protection, but for various reasons these layers of protection were rendered inoperable and no one seemed to have noticed. When the chips were down and the protection was needed, it was not available. No amount of attention to whether the plant operating teams were wearing the correct personal protective equipment or traditional 'slips, trips and falls' auditing would have prevented this occurrence. When dealing with operations that have the potential to cause fatalities to the general public as a result of the loss of containment of hazardous materials or energy, we need to have a much more specialised form of auditing carried out to ensure that potentially hazardous operations continue to be conducted safely. In these circumstances, it is necessary to have systems in place that will provide assurance that the process safety is adequate. What is even more important is to understand that there is no simple panacea in process design. Mitigation controls are designed with a specific purpose and always have limitations. It is crucially important that operating companies understand the limits or shortcomings in their process designs.

SO WHAT DO WE MEAN BY 'PROCESS SAFETY'?

It is *a discipline that focuses on the prevention of physical situations which have the potential to cause human injury, damage to property or damage to the environment through the release of chemical energy in the form of*

- *Fire*
- *Explosion*
- *Toxicity*
- *Corrosivity*

The use of the term 'process safety' can be confusing in that often organisations that operate outside of the 'chemical process' industry think that it cannot apply to them. It is important to take the widest possible interpretation of the term 'process'. It should be remembered that many municipal swimming pools and hospitals have enough chlorine on-site to cause multiple fatalities. The major explosion in Toulouse in France in September 2001 which left a crater in the ground 300 metres in diameter was not involved in a chemical manufacturing process; it was just storing fertiliser! Some soft drinks manufacturers use large quantities of acid and alkali in the cleaning of their canning lines, and likewise, farmers can store and handle significant quantities of pesticides and acids and breweries can produce large quantities of carbon dioxide. Process safety hazards depend on the nature of the work being done and the materials being handled.

Wherever process safety risks exist, these must be controlled to a tolerable level. The first requirement is to make the process that is being used as intrinsically safe as possible. That means safe by design. In many cases, it is not possible to eliminate the risk altogether. For example, if your organisation produces or uses explosives, then there are inherent risks with the product. So long as you remain in that business, you can mitigate the risk so far as is reasonably practicable, but there will always be some residual risk.

When organisations select or design a process to manufacture, store or reprocess a material, it is normal to carry out a process hazard analysis (PHA). This may be in the form of a risk assessment or a much more detailed hazard and operability study. The purpose of this PHA is to ensure that all aspects of the process operation have been considered. In particular, these processes examine not only the normal operation but also abnormal operations such as start-up or shut-down and upsets. They also examine the foreseeable misuse of equipment. The learning from these very extensive studies are checked against the design, and where omissions are found, the design is amended.

However, as we have seen at Bhopal, the integrity of the design is not the only thing that can affect the safe operation of a process. Even the most sophisticated process, such as a nuclear power plant, has to be operated by people, and human beings are prone to make mistakes. Human errors can occur in all the three elements of the good EHS system that we discussed in Chapter 1. Plant and equipment are designed by people, and so we must assume that the design is probably not perfect. The procedures and instructions that are the rules by which the organisation operates are written by people, and people have limited experience and can make mistakes in the writing of these procedures. And finally, of course, people manage, operate and maintain the plant and equipment that is used, and in doing so they may forget to do things, misunderstand instructions or even occasionally wilfully violate their instructions. Process safety is always multi-faceted, and because it is recognised that things can go wrong, process safety has to be assured by multiple layers of protection, such that if one layer fails, the consequences are not severe. I liken this to the layers of an onion, so that each protective skin is reinforcing others (Figure 30.4).

There is a tendency to consider process safety to be a purely technical issue. Many operators of potentially hazardous processes utilise highly automated computerised control systems and adopt the process safety philosophy that the hardware automation will prevent anything from going wrong.

Let us consider the sequence of events that led up to a hazardous event occurring. In any process, unplanned upsets or excursions will occur from time to time.

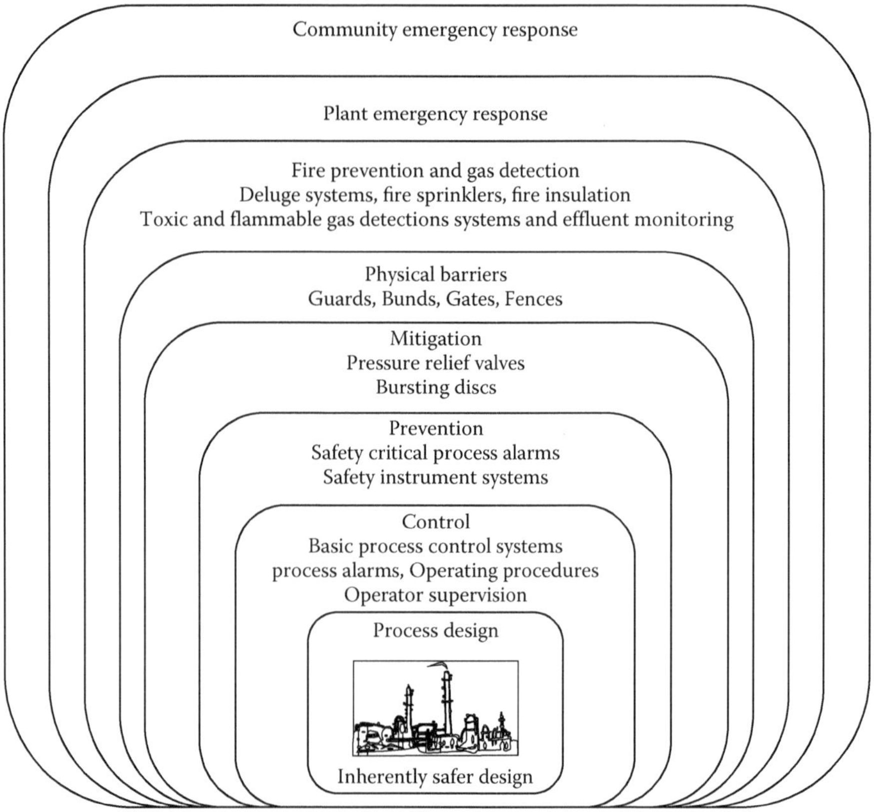

FIGURE 30.4 Typical process safety layers of protection.

Following the Venn diagram in Figure 30.5, we can see that for the situation to escalate, it requires the manual or automatic controls to fail to intervene to remedy the upset.

This failure in intervention can occur because either the upset was not detected or because the human response to detection failed to remedy the upset. Likewise with automated or passive protection systems, if the alarms do not function correctly or the automatic controls fail to intervene in an adequate way, then the automated protection systems will fail. A hazardous event will occur when both the process upset escalates and the protection systems either don't exist or are not functioning. Understanding these complex human and technical controls is a challenging task for the auditor and shows why in process safety specialist audits it is necessary to use auditors who are familiar with these types of situations.

Management who are responsible for the safe operation of their organisation must always have at the back of their mind, 'What could happen if things go wrong?' They need to be able to assure themselves that their plant, procedures and people will react in a way that maintains the process integrity at all times and avoids a loss of containment that could result in injury to people or harm to the environment. The

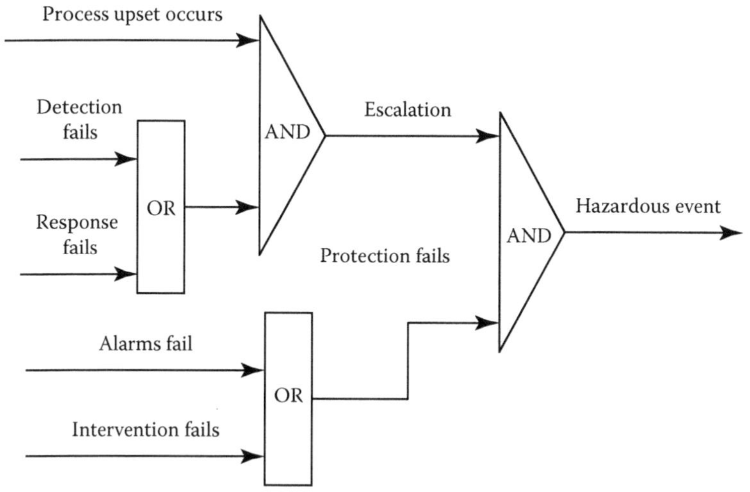

FIGURE 30.5 How a process upset could lead to a hazardous event.

way to assure yourself that you are doing all the right things and doing them in the right way is through process safety auditing.

Process safety audits are a form of Level 2 (specialist) audit. As with all specialist audits, the audit must be carried out by a specialist who is knowledgeable in the various aspects of process safety. This specialism is rarely found in one individual because it covers some technical engineering issues as well as some more traditional human safety factors. The engineering aspects also tend to be split between aspects of asset integrity (which tend to require understanding of chemical and/or mechanical engineering) and control systems integrity (which tends to require electrical/electronic engineering skills). Consequently, a team for a full process safety audit would typically comprise

- A safety specialist
- A mechanical engineer
- A control/electrical engineer

It is worth mentioning at this stage that it is not always appropriate to carry out a full process safety audit on all aspects of process safety. The need for an audit often arises as a result of some learning event where there is a need to assure the organisation about some specific aspect of process safety. For example, if during a trip-testing routine it is found that the electric actuator is missing on the control valve, it may be decided that there should be an audit carried out to establish the robustness of the trip-testing program and to ensure that it incorporates current best practice. In these circumstances, a specialist trip-testing process audit would be initiated with a limited remit. It is still advisable to make use of an auditor from outside the operating team (i.e. a 'second party' auditor) in order to gain the benefits of cross fertilisation of knowledge and add credibility to the results of the audit.

FULL PROCESS SAFETY AUDIT

The process safety audit needs to cover all aspects of operation that are necessary to ensure that the enterprise can run its potentially hazardous operations without serious incident or injury. Because some of the requirements relate to how people work and behave, there is inevitably some overlap with conventional EHS management audits. The effective operation of safe systems of work is equally as important to process integrity as it is to personal safety, and so this features in both types of audit. Generally, according to the OSHA's 'Process Safety Management: Guidelines for Compliance' (OSHA 3133), there are 12 recognised aspects to process safety management. These are

1. *Process Safety Information*
   Hazards of the Chemicals Used in the Process
   Technology of the Process
   Equipment in the Process
   Employee Involvement
2. *Process Hazard Analysis*
3. *Operating Procedures*
4. *Employee Training*
5. *Contractors*
6. *Pre-startup Safety Review*
7. *Mechanical Integrity of Equipment*
   Process Defences
   Written Procedures
   Inspection and Testing
   Quality Assurance
8. *Non-routine Work Authorisations*
9. *Managing Change*
10. *Incident Investigation*
11. *Emergency Preparedness*
12. *Compliance Audits*
    Planning
    Staffing
    Conducting the Audit
    Evaluation and Corrective Action

My experience shows that this list should be extended. The OSHA requirement for Mechanical Integrity is too limited. Process safety incidents do not only arise through mechanical equipment failures. Loss of containment can arise from any form of failure of the asset's integrity, which could be initiated by a control system malfunction, and so the auditing of process safety should extend to incorporate all aspects of Asset Integrity.

The auditing of training should also incorporate a check on competence. The key requirement for personnel is not just to be trained but to be competent. Training is purely the first step on the road to competence. To achieve full competence, it is necessary to ensure that there is validation of the training to ensure that it has been understood and remembered, and then that must be followed by some process of monitored practising of that training so that it becomes an inherent skill.

The two remaining items that I incorporate as key aspects of any process safety audit are Management Commitment and Safe Systems of Work. So my recommended list of audit aspects for a process safety audit are

1. Management Commitment
2. Process Safety Information
3. Process Hazard Analysis and Risk Assessment
4. Operating Procedures
5. Safe Systems of Work
6. Employee Training and Competence Assurance
7. Management of Contractors
8. Pre-startup Safety Review
9. Asset Integrity
10. Non-routine Work Authorisations
11. Managing Change
12. Incident Investigation
13. Emergency Preparedness
14. Compliance-Level Auditing

Process Safety Audit Aspect 1: Management Commitment

Management set the tone of the organisation. This includes setting policy, direction and priorities. The first question for any auditor is whether the current senior manager has either issued or endorsed a safety, health and environmental policy and whether that policy explicitly or implicitly covers process safety. An indication of how local management perceive the importance of the policy will be the extent to which it is displayed throughout the organisation.

The auditor will then want to understand how the management team satisfy themselves that they are meeting their legal and corporate safety requirements. It is usual to find that senior managers claim that they meet all the legal requirements, but the auditor will be interested in how they know that. Is there a system in place to track changes in legislation, and how is that knowledge implemented into the site procedures? The Management Commitment discussion will usually be one of the first discussions that the auditor holds, and so he or she will be interested to know what the management consider to be their main areas of process safety vulnerability and what actions they have in place to deal with them. For example, has the management team drawn up a full list of reasonably foreseeable major hazard scenarios, and if they have, what actions have been put in place to prevent or control these situations?

Finally, is the need for process safety expertise recognised within the organisation's structure, and is the level of resourcing appropriate? It is important to realise that the absence of numbers of people is rarely the issue when it comes to safety resources. It is a purely a matter of priorities. As one site manager said to me,

'Do you think that if we had an accident tomorrow we wouldn't find the resources to implement the learning?'

Process Safety Audit Aspect 2: Process Safety Information

Information and knowledge are paramount. There should be readily available information pertaining to the potentially hazardous chemicals and substances being stored and processed, together with the details of the process and its control philosophies in order to clearly understand the hazards that exist and the risks of those hazards leading to some unacceptable consequence. This information will probably be held in a number of different formats and locations. The first places for the process safety auditor to check are the records of chemicals held and their inventories. Suppliers material safety data sheets should be held for all substances on-site, and these materials should be labelled. The technical design of the plant should be available in some sort of plant dossier. The plant dossier will not only include up-to-date details of the process design and the engineering drawings, but will include safety studies and construction and proof-test data. On older plant and equipment the dossier may be in hard-copy format (drawings and paper records), and in more modern installations it will probably be in the form of electronic databases. Both are equally acceptable, provided that the information can be easily interrogated and understood and is up to date. The key point for the auditor is that not only does the information exist but also the people who have to use that information understand its significance. It would appear that at Bhopal, someone did not understand that the refrigeration system in the MIC storage tanks was a safety critical system, and its absence should have caused someone to ask, 'Is it still safe to continue with operations without it?' Such changes are part of the management-of-change controls that must be in place in all potentially hazardous process operations. Changes that need to be controlled are not just changes involving the introduction of new equipment, but at Bhopal, the change involving the removal of refrigerant from the MIC tank cooling system rendered that refrigeration equipment useless.

Equally, the auditor must check that the information reaches those places where it is needed. I was visiting a manufacturing facility in Spain that used a new automated packaging line to pack and palletise its product. The operator on the line believed that the safety gate interlocks stopped all downstream equipment, whereas when this was checked, it actually stopped all upstream equipment. This is critically important to understand with such equipment, where the operator safety is dependent in staying out of the fenced danger zone. He didn't understand that statistics show that if anyone goes inside the danger zone and gets injured, that injury has a one-in-four chance of being fatal. What is even worse with this type of automated equipment is that movement is controlled by limit switches and detectors. The fact that the equipment may appear to be stationary does not mean that it is safe to enter, as it is still fully powered. 'Stopped' on this type of equipment actually means 'JUST ABOUT TO START!'

One of the problems with some process safety audits is that auditors take them as an audit of systems and consequently carry out virtually all of the audit from an office. It is important to remember the three Ps and that plant and procedures alone do not make a foolproof system. It is essential to check that the human factors in the system are working because so much depends on the interpretation and implementation of the procedures. Even if that part of the process is inherently safe, there will still be human factors relating to the periodic checking of such things as vessel inspections. Simple communication may also be an issue. With the increasing globalisation of suppliers and manufacturers, plant and equipment is often produced in a different country to that in which it is subsequently used. During one audit in Japan, I found that the equipment was manufactured in Germany, and so the drawings and operating/maintenance manuals were in German. The plant was designed in England, and so the control panels were all labelled in English, but the operators were all Japanese and could not even recognise Arabic lettering. The operation was run by the operator remembering that machine width adjustment came from pressing the fourth green button from the left! Hundreds of people had been involved in the design, construction and commissioning of this plant, but no one listened to the operators' complaint that they couldn't understand the control panel or labels. Process safety is not just a technical issue; many process safety problems arise from very basic human errors. In one chemical plant that I audited, beside a newly installed red button on the control panel, someone had written in marker pen 'What does this do?' Someone else had added to the note, 'Try it and see?' Process safety must not ever resort to trial and error.

Process Safety Audit Aspect 3: Process Hazard Analysis and Risk Assessment

The process safety audit must check that during the design stage the process has been subject to a suitable and sufficient process hazard analysis. This is a technique where the manufacturing or operating process is subdivided into 'bite-sized chunks' and scrutinised by considering the possible deviations that could occur from the normal operating parameters. In large-scale or complex processes, this can be very time-consuming and is frequently on the critical path of the project process, and so sometimes the auditor may find that there has been pressure to curtail this essential safety step. It is essential that the 'process hazard analysis' (PHA) is led by a trained and competent leader who is reasonably independent of the team running the project. The quality of the PHA will depend on the knowledge and experience of those attending. Regulators are now increasingly asking 'who attended each of the PHA meetings?' and if on one occasion, say, the local Electrical Engineer was not present, then the Regulators are asking whether that part of the PHA was done competently.

Some versions of PHA like the 'hazard and operability' (HAZOP) process have up to seven tiers of checking, particularly for recognised high-hazard process operations. These tiers take the form of formal team-based analytical studies and are done at various stages throughout the project from inception through to ongoing operation. The problem is that many people do not start the process soon enough. I was asked to do a hazard study on a project for a new automated product handling line,

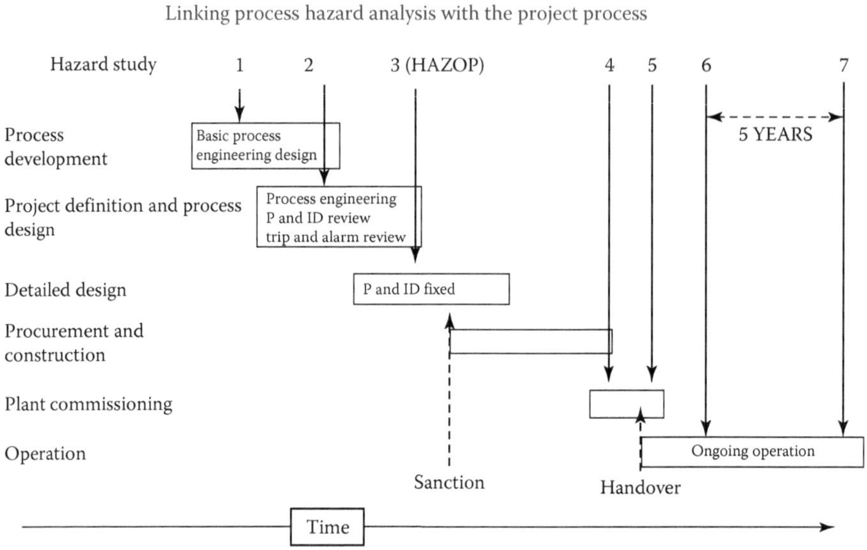

FIGURE 30.6 Timing of process hazard analyses.

only to find that it had already been installed and was about to be commissioned. Corrections and safety improvements at that stage are very expensive (Figure 30.6).

The seven recommended steps in the process are as follows:

Study 1: Review that all the information relevant to safety, health and environmental protection is available
Study 2: Identification of all the significant hazards and ensuring that the chosen design is as inherently safe as possible
Study 3: Hazard and operability (HAZOP) study
Study 4: Pre-startup safety review (PSSR) to ensure that the equipment has been installed as designed and that instructions and training are all in place
Study 5: Statutory check to ensure that the installation meets the local regulations
Study 6: Post-commissioning check to ensure that the learning from the project is recorded for the future
Study 7: Ongoing periodic review to take account of changes

The minimum number of tiers in the process is two. That is the conventional HAZOP followed by a PSSR, which does a final check to ensure that equipment has been manufactured and installed to the design and that any additional requirements identified by the PHA have been incorporated. Records of all these safety studies should be available to the auditor in the plant dossier or its equivalent. It is very common for the process safety auditor to find that some of the pre- and post-start up safety reviews were never formally carried out.

Process Safety Audit Aspect 4: Operating Procedures

A key part of operating a plant safely is how the equipment is operated. Again, this is a more of a human-factors issue than a technical issue because it involves people both in writing the instructions and also in applying them. Most competent enterprises have some sort of rules or instructions, and in many places there is a surfeit of them. There are a number of common problems that the process safety auditor needs to check.

1. Does the range of instructions/procedures cover all the foreseeable activities?
2. Are the instructions up to date?
3. Are personnel trained and validated in the application of the instructions?
4. Are the instructions enforced and followed?

To be fully compliant in all four of these straightforward requirements is much harder than you might think. The auditor is often presented on arrival with access to a large range of procedures and it is easy to be bamboozled into being impressed. Unfortunately, it is common to find flaws in the operating instructions systems.

Often the range of instructions is wrong; usually there are too few, but sometimes managers attempt to proceduralise everything, including the opening of doors, resulting in people being submerged in instructions and being unable to recognise what really matters. It is not easy to cover every eventuality, but the auditor should be on the lookout for evidence that shortcuts are being taken. I usually encourage organisations to have a formal procedure which under strict prior authorisation allows people to depart from standard practice. Provided that this is authorised at the highest level and is limited to a specific occasion, it can control those circumstances that you hadn't previously foreseen but avoids you telling your operators to 'use their initiative'. As soon as you condone deviating from the instructions, anarchy will reign!

The most common nonconformity that auditors find with instructions is that they are out of date. It is a big task to prepare instructions in the first place, and everyone is relieved when it is done and they can then get on with other things. The auditor will need to see that there is a system for the regular review of procedures and that someone is then accountable for revising and reissuing them. If there are paper copies of the procedures, I like to look at the copies that are kept on the shop floor. I usually find that if they look brand new, then no one is using them. I like to see dog-eared pages, showing that they are regularly referred to!

Once procedures have been reissued, managers tend to heave a sigh of relief that a tough task has been completed. Unfortunately, so many people forget that they have only just started. No instruction is worth anything if it is not fully implemented (Figure 30.7).

There are four steps in effectively implementing procedures. The auditor will need to sample performance at each of these steps. So, once the instruction or procedure is found to be appropriate and up to date, the next check for the auditor is to ensure that the people who are to apply the instruction have been trained and validated. 'Validation' means checking that the training has been absorbed and that the

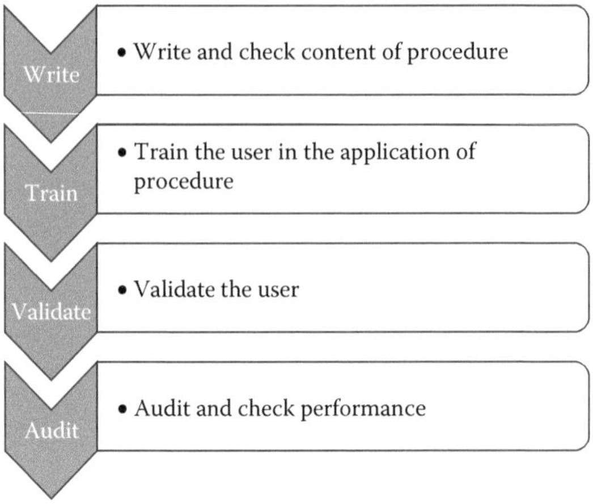

FIGURE 30.7 The steps to implementing a procedure.

person is now competent to apply that training. It is common to find that although the initial training may have been suitable, it happened many years ago and there has been no refresher training to take account of recent changes.

Finally, the auditor needs to establish that the instructions are properly enforced through adequate supervision and compliance (Level 1) auditing.

Process Safety Audit Aspect 5: Safe Systems of Work

Safe systems of work are about the totality of how work is carried out at the audited site. This topic has a direct overlap with the personal safety audits and is likely to be carried out by the safety specialist in the process safety audit team.

This part of the audit will need to examine how everyone is kept safe and without harm to their health while carrying out their work on-site. Although this extends to office workers and business travellers as well as those involved in laboratories, production and storage, the main focus for the process safety auditor will be in those areas which interface with the main potentially hazardous processing activities. Here, the auditor will need to be satisfied that the following systems are in place, understood and practiced.

- The safe control of work permits
- A system to control access into confined spaces
- A system to control the use of hot work (welding, burning, grinding etc.)
- A system to control work done at heights
- A system to control breaking into ground or structures

Details of the aspects that need to be covered in this part of the audit are detailed in Appendix A2, protocols 023, 024, 026 and 031. A very common failing in the application of safe systems of work is when the auditor discovers that the work permit system only applies to maintenance work, or what is even more common is that it applies only to maintenance contractors. The obvious inference from this is that contractors or maintenance personnel are either more valuable than everyone else or else for some reason cannot be expected to behave safely without a permit. It is imperative to understand that safe systems of work apply to everyone from the chief executive downwards, and this principle is usually enshrined within local law. The auditor should not expect that everyone works all the time under a written work permit, but the safe system of work for production personnel is likely to be the operating instructions, so it is important that these instructions are based on an adequate risk assessment and that the operating instructions do not just describe how the task should be done but identify the hazards and also any safety, health and environmental control measures that need to be applied to control the residual risks (i.e. types of personal protective equipment required to be worn and what lifting aids might be necessary to avoid back strain). When tasks are carried out for which no operating instruction exists, those tasks should be subject to a risk assessment, and where significant risks are identified, the full work control permit should be applied.

One area of work control that is particularly significant is in the use of mobile cranes working above equipment that is potentially hazardous. The fact that these machines are 'mobile' should warn users that there is a risk that the crane could either drop its load onto a sensitive piece of equipment or even more commonly that the crane itself could topple. The process safety auditor should particularly examine what controls exist to ensure that the risk of mobile cranes and plant impacting sensitive plant is effectively controlled.

Safe systems of work is one area where the auditor should expect that very frequent compliance (Level 1) auditing is carried out by the local management team itself. The auditor should verify that the compliance audits are actually happening and that there is evidence that issues are being identified and remedied.

Process Safety Audit Aspect 6: Employee Training and Competence Assurance

Virtually all regulatory regimes require that employers provide adequate training for their employees. It is easy to say, but investigation into the causes of incidents shows that one of the common root causes is failure to provide adequate training. The auditor will need to ensure that the training needs have been properly identified for all job roles and that then the job holders have been trained and validated in those tasks. This process can start with on-site informal discussions with individual workers but then needs to be verified by reference to training records. Auditors will need to check the adequacy of training by randomly checking on the records of training content and also the competence of the trainer and the content/adequacy of the training material.

It is often assumed that training at some time in the distant past is all that is needed, but things change over time and the auditor should be looking for evidence of appropriate refresher training and continuous skill development. Training alone is not what is important. As was mentioned earlier, training is just the first step on the road to competence. In process safety, individual competence is all-important, be that for the operator, maintenance technician, designer or manager. In checking competence levels, the auditor will not only be interested in evidence of adequate training but also in ensuring that the person concerned has had his knowledge and understanding of that training checked and verified in a meaningful manner. Following the validation of training, his evolution to full competence through the monitored application of the skills that have been learned is the final stage to becoming competent. Ongoing competence should then be assured through the periodic checking of those skills. In commercial aviation, pilots are trained to a very high level, but even after full qualification and many years of practical experience, even the aircraft captain is subjected to regular audit checks by one of his colleagues. The auditor should look for evidence that these audits/checks on ongoing competence are carried out and recorded.

Process Safety Audit Aspect 7: Management of Contractors and Contracted Services

The auditor needs to understand to what extent contractors or contracted services could impact on process safety. There are several areas in which contractors are commonly used which may impact on safety or environmental incidents. These are typically

1. Production operations
2. Maintenance
3. Haulage

Using contract personnel is attractive to businesses because they can often be obtained quickly without any long-term financial commitment. The danger is that those contractors may not be immediately competent to carry out work on that site. The auditor needs to check that there are systems in place to pre-qualify contracting companies to ensure that they have acceptable safety cultures and policies, experience, standards and calibre of staff to meet the organisation's needs. In the case of production operations, checks should be made to ensure that contractors are not immediately introduced to tasks without training that could lead to the loss of containment of hazardous substances or energy.

If contractors are used on the maintenance of potentially hazardous equipment, then checks should be in place to ensure that their technical work is adequately supervised by competent persons and that any necessary health monitoring is implemented.

Finally, if contractors are used for the transportation of hazardous goods, the auditor will want to be assured that all the necessary road, rail, sea and air regulations are being met and that there are arrangements in place to support foreseeable transportation emergencies. This is particularly important when it comes to the transportation of waste, where the auditor will need to confirm a chain of custody from the material leaving the site where it is produced until being received at its final destination.

Process Safety Audit Aspect 8: Pre-Startup Safety Review

The PSSR is a safety review of a new or modified processing/manufacturing plant or equipment conducted prior to commissioning, to ensure that what has been installed meets the original design or operating intent. The PSSR covers not only equipment but also procedures, documentation and training.

WHAT IS THE PURPOSE OF THE PRE-STARTUP SAFETY REVIEW?

- Ensures that the installation meets the original design and operating intent
- Ensures that adequate safety, operating, maintenance and emergency procedures exist
- Ensures that all actions from the PHA have been completed
- Ensures that training for everyone involved in the new process is completed

The auditing of some PSSR activities will have already been checked through the auditing of aspects of process safety information, training and operating procedures, but the auditor will need to assure him or herself that the essential on-site physical examination has taken place. There should be documented records maintained of these examinations and the findings arising from them.

In particular, the auditor should establish that the PSSR checks that

- All equipment identified on the drawings is present and installed correctly.
- All protective systems been correctly installed (e.g. instrumented trips, relief valves).
- All trips and alarms have been tested.
- All previous process hazard study actions been either resolved or completed (with no new hazards introduced).
- All plant operating instructions are available for all modes of operation including normal, abnormal and emergency conditions.
- All personnel have received appropriate training.
- All equipment and procedures necessary to protect the environment and for monitoring environmental performance are in place.
- All plant-based safety equipment is in place (e.g. showers and eyewash stations).
- There is adequate access for operations and maintenance.
- Fire-fighting equipment, such as hoses and extinguishers, are in place.

Auditor observations are particularly important in assessing the adequacy of the PSSR. In an audit of a brand new plant that was about to be commissioned, I noticed that the aluminium cladding on some piping insulation was partially crushed. Further investigation showed that there was a critical valve 2 metres above the pipe. It was obvious that the operator was having to stand on the pipe to reach the valve. The PSSR had not identified that safe access was required to this valve. Further drill-down demonstrated that the PSSR had failed to be done at all.

Process Safety Audit Aspect 9: Asset Integrity

Most major process safety catastrophes arise from a loss of containment of either hazardous substances or energy. It is therefore fundamental to safe operation that the equipment used is designed and built to contain those risks. The process safety auditor needs to be assured that as equipment ages and is operated and maintained that its ability to safely contain the hazards is not impaired.

The auditor should look for evidence that equipment which is critical to sustaining process safety has been identified and is subject to periodic checking by a competent person. The type of equipment that is likely to be safety critical will vary depending on the nature of the enterprise and the processes being used but is likely to include

1. Pressurised systems (e.g. pressure vessels, relief valves, some pipework)
2. Electrical systems
3. Machines that could release hazardous substances if they failed (e.g. flammable gas or toxic gas compressors)
4. Functional safety systems which have a trip function that is safety critical
5. Emergency mitigation equipment (e.g. fugitive gas-scrubbing plant, fire water drench systems, fire pumps)
6. Support structures that need to retain their strength in a fire situation
7. Bunds and drainage systems (Check that bund containment has not been compromised or drain valves left open.)

The audit checks will look for some evidence that a thorough assessment of what equipment is safety critical has been done and that a system of periodic inspection and testing is underway. The auditor will then need to review the periodic testing schedules to ensure that these are being completed on time and that any identified repairs are promptly carried out.

It should be noted that in many countries some of the asset integrity inspection programmes are also a legal requirement.

Process Safety Audit Aspect 10: Non-Routine Work Authorisations

Much of the daily work that we all do falls into the definition of routine. This means that we have done it frequently before. It is highly likely that we are very familiar and competent at doing this type of work. Very often, this is the sort of work that will be covered by documented work instructions and training. However, it is often not possible to proceduralise every single aspect of our daily work, as occasionally we will be confronted by work that we have not seen before. Nevertheless, most regulatory regimes do not differentiate between the safety requirements of routine and non-routine work; there is a legal obligation to ensure that everyone remains safe whatever they are doing. My experience is that most organisations make a reasonable attempt at controlling routine work, but many forget about the non-routine. It is quite likely that some of these non-routine jobs, such as maintenance, construction, plant trials and handling unusual or emergency situations, actually present some of the most significant hazards. So the process safety auditor needs not only to look at the permit-to-work systems and how they are applied but also to look for a formal system for dealing with abnormal situations. The auditor will normally need to look for the existence of procedures which might be entitled

- Non-routine risk assessment or job safety analysis
- Unfamiliar task assessment
- Authorisation to depart from standard practice

Even if it appears under an alternative title, the auditor is looking for evidence that the organisation has a robust system in place to ensure that risk assessments or job safety analyses are done for any of the circumstances that are covered by their definition of what constitutes a non-routine task.

Process Safety Audit Aspect 11: Management of Change

Part of the drive towards business survival entails enterprises continually trying to maintain and improve their fixed assets. Employees are often encouraged to 'use their initiatives' and solve problems for themselves. This is all well and good, except when it comes to production processes that can be potentially harmful. In the infamous explosion at Nypro Ltd at Flixborough in the UK in 1974, about 40 tonnes of cyclohexane were released from a temporary bypass pipe which was installed when one of five cascade reactors was removed for repair. The pipe modification was designed in chalk on the workshop floor and when fabricated supported on scaffolding. There was no competent professional mechanical engineer on-site at the time and the work was carried out under severe time constraints. This absence of competent persons is a key question for the auditor: how does the auditee know that people making decisions about changing designs and equipment are competent to do so? The piping modification at Flixborough was a dogleg shape mounted between two sets of expansion bellows. When the system was pressured up, no one had realised that there would be a rotational force on the pipe. This caused the bellows to rupture and release the highly flammable liquid, which vapourised into a gas cloud 100–200 metres in diameter, which then detonated.

The process safety auditor must ensure that for potentially hazardous equipment, all changes are controlled and authorised by competent persons. The checks should ensure that the changes are designed in a similarly thorough manner to the original equipment and also that pre-startup safety checks are carried out. The most common shortcoming in the application of management-of-change systems is that the final step in the process gets missed: the updating of line diagrams and engineering drawing is left until some future date, and that future date never quite seems to arrive. The problem with this oversight is that the next time a change is made to that piece of equipment, people could be working to the wrong drawings.

The other area that is commonly overlooked in management-of-change systems is its application to all of the three Ps. Like so much in process safety, the management of change is usually considered to be about changes to the hardware or computer software. Few enterprises consider the consequences of changing the third P, which is people. Organisational change can be one of the most significant changes when it comes to process safety. In his book *Lessons from Longford*, Professor Andrew Hopkins points out that one of the human factors relating to the major hydrocarbon gas cloud explosion on the No.1 Gas Plant at Longford in Australia in September 1998 was that on the day of the explosion there were no engineers or managers on-site and that the supervision was being carried out by 'deputies'. Professor Hopkins claims that the management team had been relocated from Longford to Melbourne as a result of a cost-saving measure without really considering the effect on the plant safety and operability. As has been mentioned before with respect to the Texas City explosion, on the day of the Longford incident, the plant manager was attending a safety presentation off-site. It is so easy to become complacent. The auditor must ensure that process safety considerations are taken into account when key personnel are replaced and when organisations restructure or downsize.

Process Safety Audit Aspect 12: Incident Investigation

Research carried out by BST Inc. in the United States suggests that in any one year, typically 80% of injuries will be repeats of injuries that have happened before. One could conclude from this that if you want to reduce the number of injuries that will occur in the next 12 months, the best thing that you can do is go back and look at last year's injuries and make sure that they cannot happen again. It is the concept of 'moving forwards by looking backwards'! The principle is not limited to personal injury prevention but is equally important in process safety, but in this case, you are more likely to be looking at learning from spills and leaks as well as injuries.

The key point for the auditor in checking the effectiveness of incident investigation systems is to understand what the site classifies as an incident requiring investigation. The Bird triangle shows us that if only those incidents that are reported to the regulator are investigated, then all those near-miss events at the bottom of the triangle get ignored and all those learning opportunities are missed. The auditor will need to see that every opportunity is taken to learn from what goes wrong and prevent a recurrence.

The auditor needs to seek assurance that those people who are carrying out investigations are competent to do so and are trained. In particular, the auditor should check incident reports. If the reports repeatedly say that 'Fred must be more careful next time', then this is an indication that the investigation process is not getting to the underlying or 'root' causes. Incidents invariably evolve like a jigsaw; most of the circumstances that will lead to an incident will have existed previously. It just takes someone to come along and put the last piece of the jigsaw in place for the incident to happen. Often that person will have a significant responsibility for the incident occurring at that particular time, but who was responsible for allowing the situation to arise where everything was waiting for the incident to happen? If we want to prevent not only a repeat of that specific incident but also other similar incidents, then the investigation system must identify the root cause. Frequently, root causes are failings in systems, and so the auditor should be reviewing investigation report actions to see if they address not just individual's failings but also failings in the systems. The other common failing within investigation systems is a failure to implement actions in a timely manner. The auditor should ask to review the investigation action tracking system, because the investigation alone will not result in improvement; to do that, the investigation team's recommendations need to be fully implemented.

Process Safety Audit Aspect 13: Emergency Preparedness

Thankfully, real process-related emergencies are quite rare events. The infrequency of these emergencies is actually a problem when it comes to being prepared, mainly because most people will not have experienced a genuine emergency before. Consequently, being prepared for such an event is all about anticipating the sort of things that can go wrong and then devising controls and practising over and over again. It is a natural human reaction for people to 'freeze' when something goes wrong. The desired remedial action in an emergency must become second nature. We don't want people scratching their heads to see if they can come up with a good idea while the plant is in flames around them. In checking how prepared a facility is to deal with emergencies, the auditor will need to assure him- or herself that the site has

1. Identified all the process hazards
2. Identified foreseeable emergency scenarios
3. Identified actions to deal with those scenarios
4. Taken action to ensure people who might be affected are removed from the risk (alarm systems and personnel evacuation)
5. Identified mitigation measures (emergency containment, impounding basins, gas scrubbers)
6. Provided fire control measures that are suitable and sufficient and in a state of continual readiness
7. Established appropriate links with the external emergency services

The first three items on the list are part of the emergency 'pre-plan' and should be in a format that any auditor can request to see. The role of the auditor is not to judge whether the scenarios that the facility has chosen are the correct ones or not but to be assured that the scenarios exist and have been arrived at by a systematic and thorough study. The precise type and location of fire control equipment is the subject of a specialist fire safety audit, which is sometimes carried out by the external fire authority, but the assurance that the equipment is available for use and likely to work on demand is the role of the process safety auditor. This will entail checking that there are processes in place to inspect and maintain fire equipment and that those inspections are all up to date.

It must be remembered that it is not only the large companies handling bulk quantities of hazardous materials that need to be prepared to handle emergencies and crises. There are many small organisations that handle enough quantities of hazardous materials for things to go wrong and affect the public. The author's unique SHEEMS emergency management system has been designed as an aid to small organisations and schools handling emergency situations (see Figure 30.8 or go to www.solway-consulting.com).

FIGURE 30.8 SHEEMS emergency management system for small and medium-sized enterprises.

Process Safety Audit Aspect 14: Compliance-Level Auditing

The eminent American quality guru W. Edwards Deming showed the world the importance of the 'plan–do–check–act' model (see Chapter 32), which is the basis of all modern quality systems, including those international standards that relate to auditing, safety management and environmental management. The 'check' step in Deming's system is the audit. It is essential that the organisation does not rely on occasional process safety audits by second or third parties to assure themselves that they comply. So the process safety auditor needs to look for evidence that there is robust Level 1 (compliance) auditing in place, which is administered and conducted internally within the local organisation. The auditors need to check that

- There is an audit plan in place.
- The plan is being met (i.e. audits are carried out when scheduled).
- The auditors are trained and competent.
- There is a system for dealing with corrective actions.
- The corrective actions are being completed in a timely manner.

31 EHS Aspects of Due Diligence Audits

There can be a number of reasons for undertaking due diligence audits; this may include property acquisition or divestment, refinancing, facility closure or bankruptcy, but most commonly 'due diligence' is carried out when one organisation is considering the purchase or takeover of another in the form of an acquisition. It is the application of the principle of *caveat emptor* or 'buyer beware' and is intended to ensure, so far as is reasonable, that the purchaser really understands what it is that they are buying. The major part of any due diligence process involves the organisation's financials because that is the whole *raison d'être* of any acquisition. Financial information is generally in the public domain, and so accurate information is relatively easy to find, but often information about the environmental condition and history of a location may not even exist. Although the EHS component of a due diligence exercise is a relatively minor part of the overall scope, the consequences of getting it wrong are among the most far-reaching. It should be remembered that some financial institutions have specific requirements for environmental due diligence before they will offer financial support or loans.

It must be recognised that the objective of carrying out due diligence audits differs markedly from the typical EHS audit. The overriding objective of a conventional environmental, health and safety audit is to confirm that standards are being applied and practised. This is not the case for due diligence, where the overriding objective in EHS terms is to identify what risks and liabilities could come along with the purchase, and if necessary to identify the financial consequences of those liabilities. There are many examples where environmental, health and safety liabilities have been so great as to either devalue the cost of the deal or to become a total showstopper. There are also many cases where better due diligence auditing should have prevented a contract being signed, resulting in senior executives ruing the day that they ever went ahead with the deal. This is especially the case where western companies are currently rushing headlong into diversifications in some Second- and Third-World countries without proper due diligence, only to find that they are acquiring a lot of hidden (or not-so-hidden) problems which can undermine the long-term viability of that operation.

Due diligence is primarily aimed at identifying liabilities. The problem is that this is almost always carried out in great haste, because these transactions are 'stock market sensitive'. There is never enough time to do all the due diligence auditing that a cautious buyer might like, and so it is necessary to address priorities and to initially audit those areas where there are most likely to be significant issues leading to unacceptably big financial liabilities. This is actually more difficult than it sounds, because almost invariably when such deals are being considered, share prices can be significantly affected. The possibility of 'insider trading' is taken very seriously by the authorities, and so to maintain confidentiality, the number of people involved at the early stages is usually very small. This means that the normal audit practice of talking to as many people as possible cannot apply. Once the go-ahead is given to carry out 'due diligence' EHS auditing, the audit needs to be planned and managed very carefully. It should be recognised at the outset that there are two objectives from the environmental, health and safety auditing process and these fall into the usual categories of 'benefits' and 'disadvantages'.

The environmental, health and safety benefits of the acquisition will tend to fall into the areas of such things as ISO standard accreditations, technical best practices and the organisation's culture. These benefits are often desirable, but unless the purchaser is desperate to improve their culture or to acquire some special environmental technology, these sorts of benefits are rarely deal makers. On the other hand, environmental, health and safety issues can lead to potentially large and hidden financial liabilities. And so, if time is limited, due diligence EHS auditing priority should initially be focused on identifying significant EHS-related liabilities.

In carrying out the EHS due diligence audit, the auditor will typically need to review the following information as a minimum:

- The EHS policy of the organisation
- The EHS management system
- A hazard register or equivalent
- EHS procedures and instructions
- List of hazardous materials in use
- A description of the risk management system
- Compliance with licences and authorisations
- Records of injuries and incidents
- EHS responsibilities/organisation chart
- Reports to and from regulatory bodies
- Employee handbook
- Safety training records
- Ground and groundwater investigation reports
- Site histories
- Environmental reports
- Waste disposal arrangements
- Current and potential litigation
- List of insurance claims
- Emergency management plan
- Fire management arrangements

If the organisation has significant process-related risks (see Chapter 30), then the following should be added to the list.

- Process flow diagrams
- Plant dossiers
- Process hazard assessment records

- Records of periodic testing of safety critical systems
- Records of control of design and other changes
- Maintenance records and evidence of the periodic integrity testing of critical equipment

This list should be considered to be a basic starting point for the EHS due diligence auditor. It may need adapting and extending, depending on the nature, diversity, risk and location of the proposed acquisition.

SAFETY COMPONENT OF DUE DILIGENCE AUDITING

Many of my professional occupational hygiene colleagues frequently complain that safety, health and environmental management in organisations is actually 'She' – that is, safety with a capital 'S', environment with a small 'e' and occupational health with an even smaller 'h', because managers don't really understand that occupational health management is not about just handing out aspirins when someone has a headache! In most organisations, safety is the number-one priority out of the three disciplines. If there is any good news to be had about a safety injury, then it is that provided an injury is not instantaneously fatal, then the injured person statistically has a very good chance of making a full recovery. This is not the case with harm to workers' health or the environment. In both these cases, the causes may be 'chronic' and progress over a very long period of time, but very often there is a long-term residual problem left behind. Dealing with these long-term problems costs money and is a potential major liability to the organisation concerned. However, before considering the two major areas of potential liability, there is an area of safety due diligence which requires the auditor's attention.

At the outset of the audit, the auditor must clearly identify the hazards and risks associated with the organisation. In some cases, these hazards will be physical hazards, which are typical in construction, service and agricultural industries. In other cases, the hazards may be chemical or biological, electrical or high or low temperatures, or even ionising radiation. In practice, the hazards in any one organisation are often a combination of all these. It is only when the auditor understands the hazards and risks existing within the organisation that he can drill down into the liabilities. It is always recommended that, as mentioned in Chapter 9, understanding the risks in an organisation can only be fully appreciated by undertaking a fully representative tour of the facility. Where some due diligence falls down is in making the assumption that two apparently identical facilities will have all the same risks and hence the liabilities from one facility can be extrapolated to the other without a physical visit. This is rarely the case.

Once the hazards and risks are understood, the auditor needs to study all the recent recorded injuries and incidents to establish whether there are ongoing costs associated with them. If there have been fatal accidents, then the auditor must establish whether court action is pending, which could lead to either fines or damages being levied against the new owner. If the injuries are less severe, the auditor should check if there are likely to be worker compensation claims to pay or whether the company has any voluntary schemes for providing physiotherapy or trauma counselling in order to allow the injured person to resume normal work.

Finally, in considering safety-related liabilities, the auditor needs to review the preventative actions arising from the investigations into safety injuries and incidents to assess whether there are any high-cost actions still outstanding. In particular, the auditor should be on the lookout for retraining commitments where they could apply to large numbers of people or the need to prepare new procedures and instructions, as these can often also have a misunderstood commitment for retraining or the provision of new tools or equipment.

Finally, one of the highest cost commitments in the safety area are fires. The auditor should check whether there have been any significant fires in the recent past and to what extent the damage to property and assets has been remediated. Special attention should be paid if the fires have occurred in areas containing flammable or combustible materials, and also if there have been a series of minor fire outbreaks, suggesting that the new owner has a culture of poor fire management and that further fires could be expected in the future. During the site tour, the auditor should pay particular attention to the fire mitigation and control measures that are physically in place.

HEALTH COMPONENT OF DUE DILIGENCE AUDITING

As mentioned previously, many managers tend to focus primarily on safety improvement because they can see and understand injuries. Contrary to many people's expectations, reparation for safety injuries in westernised countries is not the larger cause of financial compensation. Failure of health due to work-related causes is by far the more significant cost in compensation terms. Many managers struggle to understand the concept of occupational health as they don't have a direct parallel in their own private lives. Perhaps the closest concept is the one of ongoing 'wellness' or the prevention of bodily harm. Usually, occupational harm occurs progressively as a result of some ongoing exposure to hazard and is often not as a result of a single acute event. The classic example is noise-induced hearing loss. On the first occasion that someone is exposed to high noise, they may notice a short-term loss of hearing, but after a period of time their hearing appears to return to normal. After each exposure, more hair follicles in the ear become damaged and slowly less and less hearing returns. I started my career in a steel works and bizarrely in those days it was a matter of pride to be deaf, as it showed that you had worked there a long time and were therefore very experienced! Many people joke about deafness, but it can be life-changing for the person affected. Very frequently, as deafness progresses, the affected person finds it difficult to be in noisy public places like bars or football stadiums because they find it difficult to hear what other people are saying. Suddenly, they find that life becomes easier by staying at home. All of a sudden, the apparent joke of deafness has resulted in a total loss of their social life and the 'fun' part of life comes to a premature end.

The most common cause of worker physical disablement is back pain. Very often, this is related to poor manual handling practices, where repeated lifting, twisting and stretching results in a back injury. This sort of harm is often permanent and is frequently the cause of workers having to cease employment at a younger age that they had intended. Such injury can also significantly affect the individual's leisure activities as well as work. When people have to give up work early, or their working ability is adversely affected and the cause is shown to be work related, then this usually results in a claim against the organisation for financial compensation. These claims can often be quite substantial and are not always covered by insurance. Even when there is insurance cover, repeated claims can result in large increases in insurance premiums. It can be seen, therefore, that poor occupational health management at a facility which is under threat of acquisition can result in a series of potential future compensation claims for conditions that were initiated in the past but which have not yet reached the severity whereby the individual affected is unable to work. Particular care must be exercised when acquiring organisations that have previously gone into receivership, since the previous owner may no longer exist or be financially able to pay their share of occupational health-related compensation claims, and so the liability falls back onto the current owner.

Certain operations are particularly susceptible to occupational health claims. Those workers who are exposed to working with lead, arsenic, asbestos, dusts and repeated manual handling or noise, to name but a few, are partially protected by regulation, but the due diligence auditor must be vigilant in looking for historic examples of occupational health failures and claims. A discussion with the company's occupational physician or nurse may be a good starting point for the auditor.

It is also important to check how extensive the records are relating to training and the exposure of individuals to health-harming situations. In order to protect the organisation against future claims for industrial deafness, for example, it may be necessary to be able to prove that

1. The individual was trained on a particular date about what action to take to prevent hearing loss.
2. The individual was issued with hearing protection on various specific and recorded dates.
3. The individual was reprimanded on specific occasions for failure to wear hearing protection.
4. Records show that the individual only worked for x% of his time in noisy areas; the rest of his employment with this organisation he worked in quiet areas.
5. Pre-employment audiometry testing showed that he already had substantial hearing loss when he joined the company.

Some of this type of information is typically found among the human resources department's files, and some may be with the company medical department. The auditor will not be expected to prove whether every future claim will result in a compensation award but rather to observe whether there are systems in place that will enable the buyer to counter any unreasonable occupational health compensation claims. The auditor will of course need to identify any current occupational safety and health claims that have been submitted but may not reach a conclusion before the planned acquisition date.

ENVIRONMENTAL COMPONENT OF DUE DILIGENCE AUDITING

The booby trap of many acquisitions is not so much the health and safety liabilities associated with the deal but the environmental legacy. The problem is that if the ground or groundwater has been contaminated, the time and cost for remediation can be very substantial. Nearly every industrial application has the potential to leave traces of its existence, and the older the facility, the more likely this is to be the case. In fact, not only industrial processes but vehicle workshops, launderettes, domestic oil storage tanks, large areas of tarmac – in fact almost anything – can leave behind a chemical footprint with traces of its presence.

Environmental due diligence is playing an increasingly important role in mergers and acquisitions. Unfortunately, contracting parties often fail to do sufficient environmental due diligence or do not complete it early enough to make effective use of the information in moderating or cancelling the transaction. Much of environmental due diligence is now enshrined in law which can impose significant liabilities on a wide range of organisations, including successor and parent companies. For example, under the Comprehensive Environmental Response, Compensation and Liability Act (CERCLA) of the United States, commonly referred to as 'Superfund', organisations may be responsible for the clean-up of contamination at facilities they currently or formerly owned or operated, as well as at disposal facilities where their waste was sent. This responsibility for not only the operating sites but also the waste disposal facilities hugely broadens the scope of environmental liability and may require potential acquirers to be responsible for remediating waste at a disposal site for which they were not initially responsible and about which they have very little knowledge. Exploring the links to past and present waste disposal sites is a key role for the due diligence auditor.

When carrying out ground and groundwater investigations, it is quite usual for the buyer to want quantitative information provided by site investigations. This information can help in identifying the scale and spread of any contamination. Identifying the scale of the problem is beneficial to the purchaser in that it

- Sets a baseline as to what contribution the new owner might make to the contamination in the future and what was pre-existing
- Allows for estimating remediation costs
- Allows for the allocation of the costs among the contracted parties

Carrying out appropriate environmental due diligence in some countries can assist with minimising future risk of prosecution. For example, under the US CERCLA legislation, this can help to establish the 'innocent purchaser's' defence.

It may be that the vendor has already commenced a programme of remediation, in which case the auditor will need to establish how significant the cost of running the remediation processes is and for how many years that will need to continue.

Identifying what constitutes appropriate environmental due diligence is not entirely clear and provides a challenge for the auditor. It is usual to expect investigations to come in several steps. CERCLA identifies two steps as Phase I and Phase II, but other investigation standards identify up to four steps. Whatever regulatory regime is applicable, the underlying principles are similar. The auditor must establish that there has been a thorough historical review carried out. This historical review may also go under the name of an environmental site assessment (ESA). This first phase does not entail any physical sampling of the ground or groundwater but is aimed at establishing sufficient information about the site to identify the following:

- What potential contaminants exist?
- In what quantities?
- Where might they be migrating to?
- What is the underlying geology?
- Who or what could be affected?

There is no point in drilling deep, expensive boreholes to sample groundwater if there is no aquifer in that strata. Much of this historical review will have been carried out by use of questionnaires and by reviewing old documents, photographs and maps, and talking to neighbours and retired employees to see what they remember happening years ago. Some of the information may be held by local government officials and regulatory bodies. It is not the role of the due diligence auditor to carry out a historical review if none has been done. He or she should identify that no review has been carried out and ask for one to be done. The conducting of these historical reviews is often carried out by specialised environmental consultants. A standardised approach to this historical review has been produced by the American Society for Testing and Materials (ASTM) and can form a useful approach. The standard is in two parts. The first part is the 'transition screen'. This entails the user to establish basic historical information about potential contamination and utilises questionnaires for the owner of the site, accompanied by a site familiarisation visit and records searches. This document is known as ASTM publication E 1528. ASTM have produced a second standard, the Phase 1 Environmental Site Assessment (E 1527-05), which details a more extensive investigation. There are limitations in using the ASTM standards approach, as these do not require the information search to go beyond those records which are held by the owner or on public record. In considering the value of a historical review, the due diligence auditor must consider that groundwater migration is no respecter of surface fencelines. It is highly likely if contamination has occurred that it has spread below ground level either into adjoining property or even from it. A particular point for the auditor to address is that if it is suspected that contamination is incoming into the site, is it likely that the source can be proven and, more importantly, is the polluter in a position to pay? It is not uncommon for some fly-by-night organisations to buy up old properties, not worry about creating pollution and then find themselves bankrupt before anything can be done about it. In most areas of the world, there is no independent source of funding to undertake the clean-up. This neighbouring incoming contamination effect is one often overlooked by due diligence auditors.

Off-site effects are also important when the site is located over an aquifer. The auditor must check whether there are any authorisations for groundwater abstraction, especially if this is for potable purposes, irrigation or applications linked to the food chain. To understand the zone of influence of the contamination, the historical review should have reviewed the hydrogeological maps for the area to identify likely underground hydraulic gradients and likely boundaries. The auditor should recognise that surface topography is not necessarily a good indicator of groundwater flow directions and that shallow groundwater may flow in different directions to the deep aquifer. This sort of detailed understanding of the hydrogeology is especially important if the site in question has abstraction boreholes for its own water supply use, because those boreholes may be attracting contaminants from considerable distance off-site. The due diligence auditor should also check the quantities of water being abstracted from the site, especially if the facility is a big water user. There have been a few cases where the abstraction rate was so great that either the underground water sources were depleted or in extreme cases they caused ground movements which have affected the stability of buildings. In these cases, the liabilities were very substantial and unexpected.

The due diligence auditor should always check if there is boundary monitoring of groundwater flows at the site and examine the records of that monitoring programme.

Of particular interest to the due diligence auditor at this stage will be to examine the history and integrity of the underground liquid services on the site, particularly if this is old and carries potential ground or groundwater pollutants. Attention should be paid to drains and sewers, underground storage tanks, pits, sumps and underground liquid transfer lines. Even if the drainage system does not handle pollutants, significant drainage leaks can lead to costly voids, sink holes and foundation instability. It is particularly useful for the auditor to ask to see any CCTV surveys of underground sewers and containers.

If the historical review identifies areas of concern, the due diligence auditor should request a site investigation. The site investigation should be tailored to address the potential contamination identified in the Phase I historical review and may involve soil sampling, metal detection, ground-penetrating radar, sinking-monitoring boreholes (known as piezometers) and many other techniques. This part of the process under CERCLA is known as Phase II. Site investigations can be very expensive and if not properly targeted can sometimes reveal little valuable information. My recommendation is that the Phase II site investigation be separated into two parts. Initially, there should be an initial screening process that allows the more detailed investigation to be more precisely located. This may use techniques such as organic gas monitoring to check for the presence of hydrocarbons or the use of simple portable ground insertion probes. Techniques at the screening stage which give relatively immediate measurements, instead of sending samples away to a laboratory and waiting for many days for a result, will be the most cost-effective at this stage.

Once the screening has located the general area of ground or groundwater contamination, a detailed localised investigation can start. This may involve digging trial pits for ground contamination investigation and sinking boreholes for groundwater investigations. Often, this work is requested and paid for by the acquiring party in the deal, but sometimes the vendor may be proactive in initiating and paying for the work. In these circumstances, the due diligence auditor must request split samples for their own analysis in the event of there being doubts about the vendor's analysis results. It is always essential to use qualified and accredited laboratories for analysis and also ensure that samples are suitably packaged. Evaporation and drying can occur in soil samples that are suspected of containing volatile compounds, leading to an unrepresentatively low concentration of contamination. Since sample testing can potentially identify expensive liabilities, the security of samples in transit is paramount. There must be a reliable and recorded chain of custody to prove that the sample that was taken was the same sample that was analysed in the laboratory.

The final stage of dealing with any contamination is that of remediation. This is usually the most expensive and time-consuming phase of an environmental clean-up operation and can often take many years to complete. It is very rare that the remediation process takes part during due diligence. However, in order to determine the acquired liabilities, the auditor is likely to have to establish how remediation might be undertaken together with an order of cost for that remediation and its likelihood of success. Sometimes, in order to close an acquisition deal, it is necessary for the vendor to retain some or all of the environmental liabilities. In these circumstances, the due diligence team may need to seek guarantees or financial bonds to be assured that the vendor will remain financially solvent long enough to carry out these responsibilities.

So far, I have focused on the biggest issues of ground and groundwater contamination. However, the due diligence auditor should also address what other contamination issues might exist on the acquisition site. This involves a physical tour of the facility. I always ask to see redundant or idle equipment, especially tanks and pipelines, as these may still contain residual products which may be difficult to dispose of. It may be necessary to sample materials from these areas. I would also always ask for samples from the sludge in large evaporative cooling tower basins and effluent treatment ponds and lagoons, as these can harbour some nasty surprises. Other places to focus for evidence of contamination include old tipping/disposal sites, road/rail tanker loading bays and damaged or unsealed storage tank bunds (Figure 31.1).

On sites which are more than 50 years old, the auditor should also expect to find evidence of asbestos in piping insulation, ceiling tiles, brake shoes and, in Europe particularly, in cement sheets. With the exception of asbestos, which could leave a

FIGURE 31.1 The auditor needs to be alert for environmental legacy issues.

legacy of as yet undiagnosed asbestosis or mesothelioma, the scale of the liabilities for most of these latter issues is unlikely to be deal breaking.

Very occasionally, environmental legacy issues can have business benefit. One company that I know of was the subject of a hostile takeover bid. The company had a history on some of its facilities that went back over 100 years, and many of these sites had unknown environmental legacies. As a defence against the hostile takeover, the company publicly emphasised the scale of the environmental legacy and eventually the hostile acquirer withdrew.

Finally, to summarise the main areas of focus for the environmental due diligence auditor, he or she will need to examine the following areas of potential environmental legacy:

1. Current use of the property
2. Historic usage of the property
3. Current and historic usage of neighbouring properties
4. Hydrogeological map of the area
5. Geological conditions
6. Descriptions of structures
7. Roads and railways
8. Water supply: piped potable and groundwater
9. Drainage systems (foul and chemical)
10. Hazardous substances list
11. Biological substances
12. Underground storage tanks, sumps and pits
13. Aboveground storage
14. Evidence of spillages
15. Drum storage areas
16. Odours
17. Vegetation damage
18. Septic tanks and effluent treatment systems
19. Ponds and lagoons
20. Wells and boreholes

32 International EHS Auditing Standards

The International Organisation for Standardisation is a federation of national standards organisations that agrees and specifies international standards in a wide range of applications. In this chapter, we will summarise the quality, environmental and safety management standards and their interrelationships.

The basis for these standards are the range of quality standards, of which the key one is ISO 9001 ('Quality Management Systems Requirements'). This standard, approved in 2015, adopts the 'plan–do–check–act' approach proposed by W. Edward Deming.

In the model shown in Figure 32.1, the four stages in the model relate to how the management system and processes are applied and implemented. An interpretation of the four stages in the model is

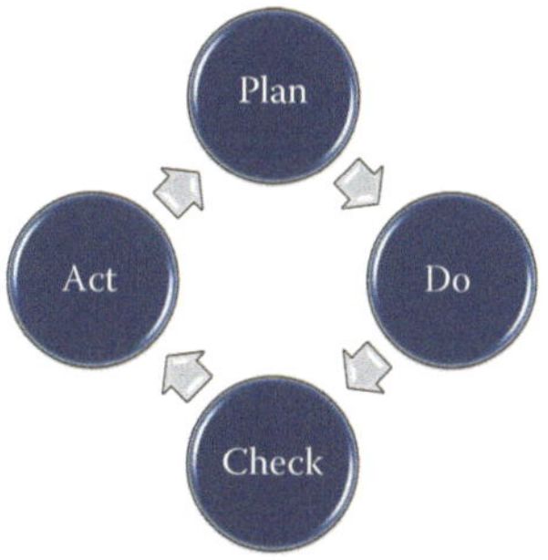

FIGURE 32.1 The PDCA model.

1. Plan: Establish the objectives and processes required to achieve the results in accordance with the organisations stated policy.
2. Do: Implement the process.
3. Check (the auditing stage): Monitor and measure the processes against the objectives set.
4. Act: Take continual improvement action.

This plan–do–check–act approach is the principle behind all three of the key ISO standards that apply to safety, health and environmental management. The circuit of using audits to identify corrective actions, which when resolved lead to improvements, then results in a never-ending upwards spiral of improvement. The application of quality management techniques to environmental, health and safety management is entirely logical, as it advocates treating the prevention of all types of harm in the same way as other aspects of managing an organisation. Why then is it necessary to have separate standards on quality, environmental management and safety, when they could all be integrated under a single quality standard? The cynics might argue that corporate registration under international standards is big business and income for the national standards bodies and their associates, and there are particular companies that successfully use the ISO 9001 quality management approach across all aspects of management, including safety, health and environmental management. It is also quite common now to find organisations integrating their environmental, health, safety and quality organisations because of the similarities in management styles.

The answer to the question is partly historical and partly to do with the way in which the quality standard is presently structured. ISO 9001 was the first of the internationally agreed standards relating to quality management. Because it is generally applicable, it needs to be relevant to a very wide range of organisations as diverse as local government, hospitals, haulage companies, financial institutions and nuclear power plants. Consequently, the standard requires organisations to establish for themselves what criteria it is important for them to meet. In the early days of the introduction of the standard, some organisations were setting very demanding requirements for themselves, while others set much less demanding ones. It led at that time to a belief that if the requirements were easy to achieve, then getting certification against ISO 9001 was not too difficult, and some felt that this gave a misleading result. There was a feeling that for some organisations, it encouraged the setting of low standards and resulted in the quality threshold being low.

The desire to apply a quality approach to environmental management opened up a new opportunity. Unlike business objectives, environmental objectives are relatively clear, in that virtually everyone lives and works in an environment surrounded by air, ground or water. The consequences of contaminating any one or more of these three aspects of the environment are well understood, and although the types of contaminants are myriad, the solutions are quite limited. This offered the opportunity for the international standard on environmental management (ISO 14001) to be much more focused and include some specified objectives. The application of ISO 14001 resulted in much less variability in authorisations, and its success has lead to a similar approach being taken to the quality management of safety in the international standard OHSAS 18001.

The use of similar approaches in the application of international quality, environmental and safety standards means that the same style of Level 1 (compliance) auditing can be universally applied. The benefit of this to the organisation is that the same auditors, auditor training, audit planning and corrective action tracking system can in theory be applied to quality, safety and environmental requirements. This has the potential to result in more competent auditors, better auditee understanding of the benefits of auditing and a single streamlined process for dealing with nonconformity and noncompliances. Most importantly, it emphasises that environmental, health and safety management is not a separate independent strand of the management process but is fully integrated within it.

A SUMMARY OF ISO 14001 REQUIREMENTS

Background

This standard aims to achieve a balance between what it describes as the three pillars of sustainability, which are the environment, society and the economy (Figure 32.2).

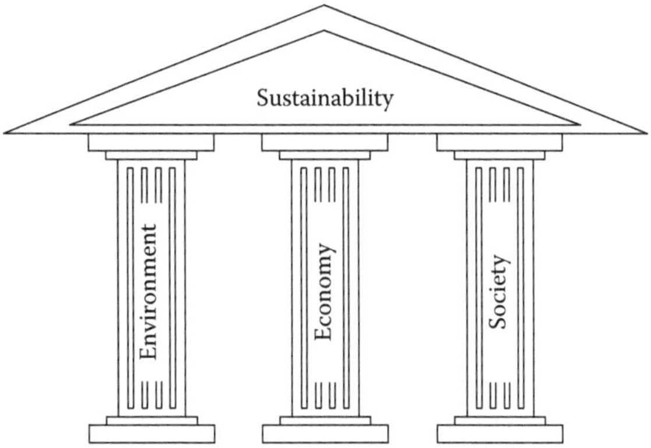

FIGURE 32.2 The three pillars of sustainability.

ISO 14001:2015 aims to provide a systematic approach to environmental management. It differs from ISO 9001 in that it does not just provide for compliance with an organisation's own standards, it also defines those areas of environmental management that are important. The standard is intended to help organisations to

- Improve environmental performance
- Achieve regulatory compliance
- Achieve the organisation's environmental objectives
- Encourage a 'life cycle' approach to the way in which products and services are produced, used and disposed of

The standard requires that the organisation has an environmental management system to control those aspects of their operation that can have potential impact on the environment. In this respect, the environment is defined as the 'surroundings in which the organisation operates, including air, water, land, natural resources, flora, fauna, humans and their interrelationships'.

To conform to the standard, the organisation has to demonstrate conformance to detailed requirements under the following categories:

1. Environmental Leadership and Commitment
2. Environmental Policy
3. Organisational Roles and Responsibilities
4. Planning
5. Support (Resources/Competence/Awareness/Communication/Information)
6. Operation

7. Environmental Performance Evaluation
8. Internal Auditing
9. Management Review

In preparing for an environmental management system, the management team must identify who the interested parties are and their requirements. This will aid the organisation in identifying for itself its environmental policy and which requirements should become its environmental objectives. In large organisations, these objectives may be tiered so that the objectives of each individual facility contribute to the achievement of the organisation's overall strategic objective.

In order to implement the environmental policy, the organisation should follow the four steps of the plan–do–check–act model described earlier. Planning requires the establishment of an environmental management system. As with any management system, an environmental management system will not exist and thrive without management's leadership and commitment.

Category 1: Leadership and Commitment

It is essential that senior management demonstrate commitment to the environmental management system through their actions. This begins by establishing the environmental policy and then demonstrating active support in the implementation of that policy by

- Setting good personal examples
- Integrating environmental objectives with the organisation's strategic direction
- Providing adequate human and financial resources
- Monitoring and adapting the systems to ensure that environmental objectives are met

A starting point for any Level 2 or Level 3 audit should always be for the senior auditor to establish the extent of senior management commitment, where the individuals concerned will be expected to demonstrate those responsibilities that they are personally accountable for.

Category 2: Environmental Policy

Senior management are responsible for the preparation and issue of the environmental policy and for communicating that policy to those affected by it. The organisation must ensure that its environmental objectives are consistent with its policy. The policy should be periodically reviewed and updated in light of relevant changes and, in particular, should be approved by the current most senior manager for the organisation or that part of the organisation.

It is important that the policy addresses commitments to the protection of the environment and the principle of 'pollution prevention'. The policy should encapsulate all those potential areas where the organisation's activities can have either a positive or negative impact on the environment.

The four principles that should be covered in the environmental policy are

1. To protect the environment
2. To comply with regulatory requirements
3. To meet the organisation's environmental objectives
4. Continuous improvement

Category 3: Organisational Roles and Responsibilities

Senior management must allocate roles and responsibilities throughout the organisation for the implementation and ongoing management of the environmental management system. This may mean the appointment of a suitably qualified and trained 'environmental manager', but more importantly, it will require the assigning of responsibilities throughout the management line and ensuring that those people at the 'shop floor' level where environmental impacts are most likely to arise are also trained and aware of the importance of their individual roles and duties.

There must be clear internal accountability for ensuring that the environmental management system conforms to ISO 14001 and that such conformity is maintained by rigorous checking in the form of Level 1 compliance audits.

The organisation must decide what environmental parameters are relevant for monitoring improvement and shall then appoint responsibilities for the gathering, analysing and reporting of that information on a regular frequency.

Category 4: Planning

The organisation must consider what parts of its operations, products or services can have an effect on the environment either at a local, regional or global level. Such parts of the operation are known as 'environmental aspects'. In identifying these environmental aspects, the organisation is required to consider not only the normal smooth running state of the operation but also

- Abnormal conditions
- Reasonably foreseeable emergencies
- New developments/products or services

The organisation must maintain documentary records of the criteria it used to identify environmental aspects and which of those it considers to be significant.

There should be a register of applicable environmental legislation kept up to date with clear accountabilities relating to the compliance assurance of each legal requirement. In addition to national legislation, there may be local binding agreements, industry sector agreements, corporate objectives or other nonnegotiable environmental obligations that need to be included when identifying which environmental aspects are significant.

The organisation must plan to address the significant environmental aspects and corrective actions to ensure compliance with regulation. It is normal that these environmental improvement plans will look forward several years into the future, as it is recognised that some environmental improvements may incur substantial capital expenditure, which requires long-term forward planning. Ideally, these plans are on a rolling basis, so that as some actions are completed, new actions are added. The plans should clearly identify who should do what and by when. Once the environmental management systems are established, the maintenance of the system is checked via the Level 1 auditing system. In order to ensure elements of the system do not get overlooked, a rolling audit schedule should be established along the lines discussed in Chapter 6 (Figure 6.1).

Category 5: Support

The organisation must decide how it will resource the establishment, implementation, maintenance and ongoing improvement of the environmental management system. To do this, the management team will need to consider financial, human, equipment and information technology (IT) resources. There is always an initial financial cost implication, even if the management system relies entirely on existing people, as there will be training, communication, systems familiarisation and debugging and new auditing commitments, all of which will take time and money. It is important that this initial cost implication is understood, as the management team's credibility will be undermined if they announce a plan to gain accreditation to ISO 14001 and then abandon it part way down the track because they don't have the resources. It should be recognised that once the system is fully implemented, there are often financial rewards because costly environmental incidents and prosecutions are avoided and often raw material and services consumption is reduced.

It is important to ensure that responsibilities are assigned to competent people, and so training and competence validation will need to be demonstrated.

A key part of any successful management system is two-way communication, and this is no different in the case of the environmental management system. The organisation must have effective internal systems to identify what it is going to communicate, to whom and when, and this information must be recorded. Most organisations have some sort of regular communication sessions that cascade throughout the organisation, and it is usual to integrate the environmental communications into these existing channels rather than create separate and parallel systems. What does not always exist is an effective means of communicating to external stakeholders, and the implementation of ISO 14001 will ensure that appropriate arrangements are made.

In common with ISO 9001, ISO 14001 requires that procedural documents are produced in a standardised format and that there is a document control process in place to authorise and issue changes.

Category 6: Operation

The organisation must have processes in place to meet the objectives set for the environmental management system. Controlling these processes may be implemented using the 'hierarchy of controls' (Figure 32.3), which in the case of ISO 14001 identifies that the elimination of the environmental risk is the best risk control option, followed by substitution by a lesser risk, and those are preferable to administrative controls, which can be subject to human error.

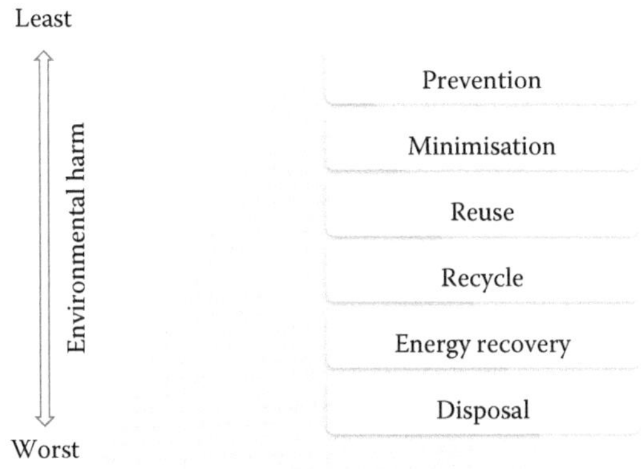

FIGURE 32.3 Environmental hierarchy.

Outsourced processes such as waste recycling and disposal also have to be effectively controlled and those controls demonstrated. These outsourced operations will need to be checked by specialist Level 2 audits to demonstrate that the controls are effective. Where outsourcing involves the use of contracted services, those contracts need to reflect the appropriate environmental objectives.

A key part of operational requirements is the need for emergency response plans, which will mitigate adverse environmental impacts from emergency situations. These situations are not just limited to environmentally related incidents but also where, for example, in fire situations there can be the release of airborne toxins such as asbestos or the escape of large quantities of contaminated firefighting water. Foreseeable emergencies should take account of severe weather and relevant product transport emergencies.

Affected personnel must be trained in the emergency action requirements, and emergency control plans must be subject to periodic testing and review.

Category 7: Environmental Performance Evaluation

Where emission monitoring and analysis is essential for effective environmental control, that monitoring equipment must be suitable and calibrated or verified as required to ensure its reliability. Records of previous monitoring must be maintained as proof of ongoing measurement, evaluation, analysis and corrective action. The checking of calibration records is a key part of the auditor's verification tasks.

Category 8: Internal Auditing

The organisation is required to provide assurance of compliance with the requirements of ISO 14001. This is done by Level 1 (compliance-level) auditing and then taking corrective action where required. An audit programme is required, and audit frequency will depend on the criticality of that aspect of the management system in prevention of pollution or harm. Local auditors need to be trained and competent, and the results of the audit must be reported and actioned by management. Documented evidence that audits have been conducted need to be maintained.

Category 9: Management Review

Senior management must periodically undertake a review of the environmental management system to ensure its suitability and effectiveness, taking into account relevant internal and external changes and the appropriateness of current resources. The review must consider the records of

- Environmental monitoring
- Nonconformities
- Corrective actions
- Audit results

Documentary records must be maintained relating to the management review.

Certification to ISO standards (previously known as 'accreditation') is carried out by a very limited number of approved certification bodies who are authorised by ISO via the local national standards bodies. It is not possible for organisations to officially self-certify conformity with the ISO standards.

A SUMMARY OF OHSAS 18001 REQUIREMENTS

At the time of writing, the Occupational Health and Safety Assessment Series standard 18001 (OHSAS 18001:2007) is the current international standard identifying requirements for occupational health and safety management systems. This standard is scheduled to be replaced in early 2018 with ISO 45001.

OHSAS 18001 was developed to compliment ISO 9001 (quality) and ISO 14001 (environmental) in response to the demand for an internationally accredited system for occupational health and safety management. The early version was launched as a specification rather than a standard, hence its designation 'OHSAS' rather than 'ISO'. The 2007 version became described as a 'standard' and the migration to ISO 45001 will complete this process, making the document fully compatible with ISO 9001 and 14001.

In common with the International Quality and Environmental Management Standards, OHSAS 18001 adopts the plan–do–check–act model described earlier, but the traditional way of portraying this to demonstrate never-ending improvement is shown in Figure 32.4.

The standard is written with the intention of being auditable, with this leading to the opportunity for formal accreditation of the organisation's occupational health and safety management system.

The current standard structure differs slightly from that in ISO 14001, but in principle it covers all the same topics. The OHSAS 18001 management system requirements are

- Occupational health and safety policy
- Planning
- Implementation
- Operation
- Communication

- Documentary control
- Checking
- Management review

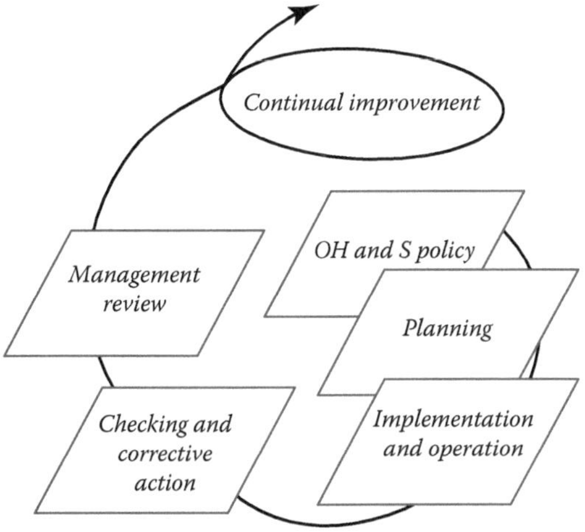

FIGURE 32.4 The OHSAS 18001 management system model. (Permission to reproduce extracts from British Standards is granted by BSI Standards Limited (BSI). No other use of this material is permitted. British Standards can be obtained in PDF or hard-copy formats from the BSI online shop: https://shop.bsigroup.com.)

Category 1: Occupational Health and Safety Policy

In many westernised countries it is now a legal requirement for organisations to have an occupational health and safety policy. For the purposes of OHSAS 18001, this policy should be suitable for the type and scale of risk that exists within the organisation's operations. The policy should be periodically reviewed and updated in the light of relevant changes, and in particular, should be approved by the current most senior manager for the organisation or part of the organisation. The policy should be communicated to all those who are affected by it.

In a similar way to ISO 14001, the four principles of commitment that should be covered in the occupational health and safety policy are

1. To aim to prevent injury and work-related ill health
2. To comply with regulatory requirements
3. To meet the organisation's occupational health and safety objectives
4. Continual improvement in the management of occupational health and safety

Category 2: Planning

The organisation must have formal arrangements in place for the ongoing identification of hazards and the consequential assessment of risks. Residual risks must be effectively controlled. These procedures should take into account everyone who can be affected by the work activity, including contractors and visitors. The risk assessment shall be proactive rather than reactive and must include

- Human factors, such as behaviour, capability and ergonomics
- Routine and non-routine tasks
- Hazards created both within the workplace and arising from or affecting adjacent work activities
- Equipment and materials at the workplace
- The design of workplaces and equipment therein

The resulting assessment must identify and prioritise risk, and where the risk is unacceptable, identify what controls are necessary to deal with the residual risk. When determining what controls are appropriate, the hierarchy of controls should be applied as shown in Figure 32.5. Control options higher up the hierarchy are considered to be more effective than those lower down.

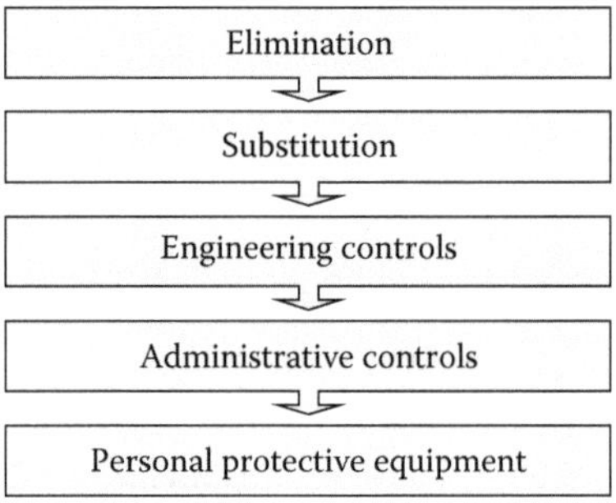

FIGURE 32.5 OHSAS 18001 'hierarchy of controls'.

The results of these risk assessments must be recorded and communicated to those affected.

As with ISO 14001, a register of applicable health and safety legislation should be maintained, with clear accountabilities relating to the compliance assurance of each statutory obligation. In addition to national legislation, there may be local binding agreements, industry sector agreements, corporate objectives or other nonnegotiable occupational health and safety obligations that need to be complied with, and elements of all these requirements may form part of the organisation's occupational health and safety objectives. These objectives must be measurable.

Category 3: Legal Compliance

The organisation must have arrangements in place to clearly identify what current laws and regulations are applicable to their operations and must ensure that these statutory obligations are taken into account in the health and safety management system, including procedures, instructions and other forms of risk control, such as guards, signs and protective equipment and so on. This information must be kept up to date.

Category 4: Objectives

The organisation must set, implement and maintain a set of written health and safety objectives and these must be in keeping with both the health and safety policy and local legislation. Performance against these objectives must be monitored. An ongoing health and safety improvement plan must identify what actions are required to achieve the objectives and who is responsible for delivering that improvement and by when.

Category 5: Roles and Resources

The senior management team are responsible for health and safety at the facility and for the health and safety management system. They are responsible for ensuring that suitable and sufficient resources are provided to allow the organisation to achieve its health and safety objectives. In order to do this, the management team must define the roles and responsibilities relating to effective health and safety management throughout the organisation. A specific member of the local senior management team must be appointed with specific responsibility for health and safety. This appointment must be communicated to all people working within the organisation.

Category 6: Training and Competence

All persons working within the facility must be educated, trained and sufficiently experienced to do their job safely and without harm to their health. In order to achieve this, the organisation must carry out an assessment of training requirements and then provide suitable training to satisfy these requirements. All persons, including employees, contractors and visitors must be informed of the hazards and risks associated with their work activities and what their own personal responsibilities may be to protect their own safety. This information should be in a form that is appropriate for the individual concerned and should take into account such things as literacy and language skills. It must be made clear to all persons carrying out work at the facility what will be the consequences of not conforming to the organisation's health and safety procedures.

Category 7: Consultation

In addition to the requirements to communicate hazards and risks, there is a requirement under the standard to ensure that workers are consulted in important decisions relating to their health and safety. This may apply to their participation in hazard identification, risk assessment, specification and choice of workplace controls and their involvement in relevant accident and incident investigations.

Category 8: Documentation

As with all ISO standards, OHSAS 18001 is required to have certain basic documentation, which includes

- Health and safety policy and associated objectives

- The scope of the health and safety management system
- Associated records (those required to ensure the system is maintained)

There must also be a document management system to ensure that both distribution and updating is effectively controlled.

Category 9: Operational Control

Control of health and safety is not just linked to workers' activities. Management controls need to be established over purchased goods, substances and services to ensure that hazards are identified and risks controlled. The management system should cover not only the controlling of acts by persons but also the consequences of human omissions.

Category 10: Emergency Preparedness

A procedure must be established for identifying foreseeable emergencies and then arrangements for dealing with those emergencies. Emergency response actions should take account of the requirements of neighbours and the emergency services. The emergency procedures must be periodically tested in order to ensure personnel are trained and lessons are learned.

A system designed to assist small and medium-sized enterprises in complying with the OHSAS 18001 requirement is the SHEEMS system, as shown in Figure 32.6 (more information at www.solwayconsulting.com).

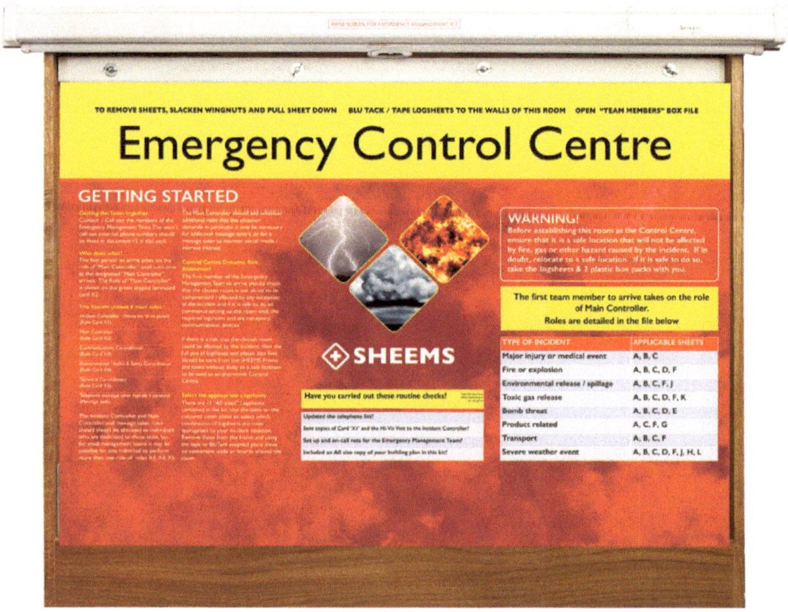

FIGURE 32.6 System for small and medium-sized enterprises that need to comply with the emergency management requirements of OHSAS 18001/ISO 45001.

Category 11: Checking (Monitoring and Auditing)

The organisation is required to provide assurance of compliance with the requirements of OHSAS 18001. This should be done by Level 1 (compliance-level) auditing and monitoring performance measures and then taking corrective action where required. A system is required to identify and respond to nonconformities, corrective actions and preventative actions. Normally, the auditor would expect to find an up-to-date action recording and tracking system.

Category 12: Incident Investigation

The organisation should have a system to record, investigate and learn from health and safety incidents that occur. In order to achieve this, the system should

- Record the incident
- Determine the underlying health and safety deficiencies (root cause)
- Identify opportunities for preventive action and continual improvement
- Communicate the results of the investigation
- Implement the preventative actions in a timely manner

Category 13: Management Review

Senior management must undertake a review of the health and safety management system periodically to ensure its suitability and effectiveness, taking into account relevant internal and external changes and the appropriateness of current resources. The review must consider the records of

- Internal Level 1 audits
- Regulatory compliance audits
- Results of worker health and safety consultations
- Health and safety performance monitoring results
- Extent to which health and safety objectives/plan have been met
- Learning from incident investigations
- Status of outstanding corrective actions
- Follow-up actions from previous management reviews
- Changing circumstances – new assets or changes in legal requirements

At the time of writing, the replacement for OHSAS 18001, which is ISO 45001, is not formally published, albeit available in draft form. The next section deals with the comparisons between the old and new standards so far as they are known at the end of 2017.

COMPARISON BETWEEN OHSAS 18001 AND ISO 45001 (DRAFT)

The new draft standard represents a sensible evolution from OHSAS 18001 and aligns with the plan–do–check–act principles and format of the other equivalent international standards, so that

- The overall intent is retained to manage the prevention of fatalities, injury and ill health.
- The focus on the role of top management in an effective OH&S management system is maintained and enhanced
- There remains a very strong focus on hazard, risk and effective control of risk.

ISO 45001 introduces a small number of requirements that are largely new when compared with OHSAS (new clause numbers are shown in brackets), including the following:

- Understanding the organisation and its context (4.1)
    - 'The organisation shall determine external and internal issues that are relevant to its purpose and objectives and that affect its ability to achieve the intended outcome(s) of its OH&S management system.'
- Understanding the needs and expectations of workers and other interested parties (4.2)
    - 'The organisation shall determine:
        a. the workers and other interested parties that are relevant to the OH&S management system;
        b. the requirements of these interested parties and which of these are added to applicable legal and other requirements.'
- Action to address risks and opportunities (6.1)
- Assessment of risks to the OH&S management system (6.1.2.2)
    - 'The organisation shall establish, implement and maintain a process for the on-going proactive identification of hazards arising in the workplace, and to workers.'
- Identification of OH&S opportunities (6.1.2.3)
    - 'The organisation shall establish, implement and maintain a process to:
        a. assess OH&S risks from the identified hazards taking into account applicable legal and other requirements, the effectiveness of existing controls and taking into consideration the hierarchy of controls;
        b. identify opportunities to eliminate or reduce OH&S risks.'
- Planning to take action (6.1.4)
    - 'The organisation shall plan:
        a. actions to address these the risks and opportunities;
        b. actions to address applicable legal and other requirements;
        c. actions to prepare for, and respond to, emergency situations;
        d. how to integrate and implement the relevant actions, including the determination and application of controls, into its OH&S management system processes;
        e. how to evaluate the effectiveness of these actions and respond accordingly.'

Consolidated Requirements

The new standard consolidates a number of existing requirements that were previously distributed across a number of clauses of OHSAS 18001 and rationalises them into stand-alone requirements.

For example:

- Management of change, which was previously referred to in five separate clauses, is now consolidated into the new section 8.2.
- Outsourcing, procurement and contractors requirements, which were previously referred to in eight separate and rather randomly dispersed clauses, are now consolidated into the new sections 8.3, 8.4 and 8.5.
- Continual improvement, which was previously referred to in six separate clauses, is now consolidated into the new section 10.2.

There are also minor alterations to the following clauses in the new draft standard ISO 45001:

- Scope (4.3)
- Leadership and Commitment (5.1)
- OH&S Policy (5.2)
- Organisational Roles, Responsibilities, Accountabilities and Authorities (5.3)
- Hazard Identification (6.1.2.1)
- OH&S Objectives (6.2.1) and Planning to Achieve (6.2.2)
- Information and Communication (7.4)
- Operational Planning and Control (8.1.1)
- Hierarchy of Controls (8.1.2)
- Outsourcing (8.3)
- Emergency Preparedness (8.6)
- Monitoring, Measurement, Analysis and Evaluation (9.1)
- Internal Audit (9.2)
- Management Review (9.3)
- Incident, Nonconformity, Corrective Action (9.1.1)
- Continual Improvement (10.2)

From and auditing point of view, with the exception of the six new requirements mentioned at the beginning of this section, the auditor should notice very little change in what is expected of him or her.

GUIDELINES FOR AUDITING MANAGEMENT SYSTEMS ISO 19011

The international standard ISO 19011:2011 has already been referred to on various occasions throughout this book. For completeness, this brief reference is included in this chapter on relevant international standards. This standard focuses primarily on setting up the audit programme and provides some useful advice for organisations doing that for the first time. That advice has been referred to and substantially expanded on throughout this book.

What this standard does not help with is detailed advice on how to conduct an efficient and effective audit and still be able to maintain cordial relationships between the auditors and auditee. However, new auditors and organisation's audit managers are recommended to become familiar with the contents of the auditing standard ISO 19011.

Glossary

accident: An unplanned event giving rise to death, ill health, injury, damage or other loss.
acute effects: Consequences are immediate.
aquifer: Underground water source.
area inspection: Process of visiting the workplace to meet people and see how work is performed and view people's behaviour and working conditions.
aspect: The SHE audit topic being studied.
assimilate: A process of reading and gaining information from documents and electronic records.
audit: A process of systematic examination to assess the extent of conformity with defined standards and recognised good practice and thereby identify opportunities for improvement.
audit checklist: A summary of the key points to which the auditor requires responses.
audit fatigue: When audits become so frequent that they start being resented or corrective action requests are ignored.
audit manager: Person appointed at the audited unit to ensure that the audit programme is fully implemented in a timely fashion. This role is primarily one of administration, and the audit manager may or may not be directly involved in the detail of the audit discussions.
audit trail: Method of confirming compliance via a paperwork route.
auditability: Ability of the auditor to draw clear conclusions regarding compliance.
auditee: Person, site or organisation being audited.
auditor: Person carrying out the audit.
behaviour: An observable act.
benchmark: Reference point.
bespoke checklist: A one-off checklist prepared by the auditor charged with carrying out the audit and derived directly from the procedure/instruction to be audited.
blacklist: A list of those items that are overdue for periodic safety inspection.
brother's keeper: Team responsibility; looking after each other.
BS 5750: An early British equivalent of ISO 9000.
BSI: British Standards Institution.
BST: American behavioural safety specialist; now part of Dekra Insight.
CAT scan: Electrical induction method of tracing underground pipes or cables.
caulking: Flexible joint between two rigid components.
caveat emptor: Legal term for 'buyer beware'.
CE mark: Declaration of conformity with safety standards (EU requirement).
CERCLA: US Comprehensive Environmental Response Compensation and Liability Act, relating to ground and groundwater contamination.
change control: Formal system to control the SHE consequences of changes to hardware, software and personnel.
chartered engineer: A fully qualified and experienced engineer recognised by the Council of Engineering Institutions.

chronic effects: Long-term consequences.
codes of practice: Practical guidance on the requirements contained in a standard. If the code of practice is 'approved', the code relates to specific legal requirements and may have a special legal status.
competency: Inherent skill displayed by an individual.
compliance: Having full conformity with a predetermined standard or requirement.
compliance audit: Audit examining the compliance with local instructions or procedures.
confined space: Any enclosed space where there is a reasonably foreseeable risk associated with that space.
conformance: Meets the requirements of a documented standard; now superseded by the term 'conformity'.
conformity: ISO definition for meeting the requirements of a documented standard
consequences: Results of an action.
continuous improvement: A process of ongoing and never-ending improvement.
convergence: Process of condensing a large number of detailed corrective actions into a manageable number of clear management recommendations.
corporate governance: Process of directing and controlling all aspects of an organisation, including the SHE performance.
corrective actions: Ways of dealing with nonconformity.
corroborated evidence: Substantiated proof.
crisis management: Managing an event which has, or could have, triggered a significant real or perceived threat to safety, health or environment or to the organisation's reputation or credibility.
dBA: A weighted decibel; a measure of sound pressure adjusted for human exposure.
***de minimus*:** Trivial.
Det Norske Veritas (DNV): Headquartered in Oslo, Norway, an international consulting firm with numerous offices worldwide. Its primary focus is on safety and environmental risk.
display screen equipment: Computer workstation.
display screen regulations: UK regulations requiring the assessment of computer workstations to control ergonomic hazards.
dosimeter: Instrument used to measure exposure to hazards such as noise or chemicals.
draw-down: The area affected by the suction from a water abstraction borehole.
drill-down: Practice of delving into more depth about a particular element or aspect during an audit.
due diligence audits: Audit carried out when company takeovers are likely, to ensure that the buyer understands what he or she is getting.
duty of care: A manager's responsibility to ensure that certain SHE protection actions are taken.
EHS: Environmental, health and safety.
element: Key requirement to ensure compliance with this SHE topic or aspect.
EMAS: Eco-management and Audit Scheme.
emission abatement: Process of reducing environmental emissions.

entry meeting: Meeting between the auditor and the auditees at the commencement of the audit.

environmental aspect: An element of an organisation's activities or products or services that interacts or can interact with the environment.

environmental effect: The consequences of an environmental release.

epidemiology: The study of how often and why diseases occur in different groups of people.

equipment integrity: Intrinsic ability of a piece of equipment to operate within its designed safety margins.

ergonomics: The applied science of equipment design, as for the workplace, intended to maximise productivity by reducing operator fatigue and discomfort.

exit meeting: Meeting between the auditor and auditees at the end of the on-site part of the audit, at which the preliminary audit findings will be shared.

exposure: The act of being subjected to a hazard (usually a health hazard).

extract: Short summary of the contents of a published technical article.

functional safety system: A functional safety system is a hardware system that detects a potentially dangerous condition and causes corrective or preventative action to be taken.

gravitas: Quality of a person with knowledge and experience and who can speak with authority on the required topic.

guidance: A documented suggestion of how a particular standard might be implemented.

HASAW: The UK Health and Safety at Work etc. Act 1974.

hazard: Potential of a substance, activity or article to cause harm.

hazard study: A structured and systematic examination of a planned or existing process or operation to identify risks.

Highway Code: Rules governing the use of the road in the United Kingdom.

HSE: Health, safety and environmental.

HSE: Health and Safety Executive; the UK health and safety regulator.

HSG: Governmental health and safety guidance notes.

human factors: SHE effects that arise from people's actions.

implementation: Act of providing a practical means for accomplishing something or carrying it into effect.

incident: Unplanned event giving rise to damage or other loss.

injury: physical bodily harm.

ISO 9000: Internationally recognised standard for business management, which ensures that businesses are operating to the same standards of meeting customers' requirements.

ISO 9001: Quality management systems requirements; a part of the ISO 9000 series.

ISO 10011: International quality assessment procedures.

ISO 14000: International standard for environmental management systems (requirements and guidance for use).

ISO 19011: International standard; guidelines for quality and/or environmental management systems auditing.

ISO 45001: The new occupational health and safety management standard due to be published in 2018 and which replaces OHSAS 18001.

ISRS: The international safety-rating scheme; a widely used commercial health and safety auditing process devised by the International Loss Control Institute.

leachate: Water which drains from a landfill site.

letter of assurance: An annual letter indicating how the unit complies with the company's standards and instructions.

Level 1 audit: Audit examining the compliance with local instructions or procedures.

Level 2 audit: Audit examining a single topic in great depth (e.g. environmental audit or electrical safety audit).

Level 3 audit: Audit done at the 'strategic' level in order to examine the adequacy of arrangements for managing safety health and environmental affairs in an organisation.

Lloyds: Lloyds of London, insurance underwriters.

local exhaust ventilation: Air mover systems that are permanently installed with the objective of reducing worker exposure to hazardous fumes.

loss prevention: Ensuring that uncontrolled losses do not occur.

management audit: Audits done at the 'strategic' level in order to examine the adequacy of arrangements for managing safety, health and environmental affairs in an organisation.

manual handling: Processes that involve the lifting or moving of items using only the power of the human body.

mentoring: Advice and coaching from a knowledgeable and trusted colleague.

MHSR: UK Management of Health and Safety Regulations 1998, which specify the requirements for a risk-based approach to health and safety management.

MSDS: Material safety data sheet; a form containing data regarding the properties of a particular substance.

NAMAS: National Measurement Accreditation Service; providing measurement and testing accreditation for laboratories to ISO 17025.

nonconformity: ISO term for circumstances that do not meet the requirements of a documented standard.

noncompliances: Actions that do not meet the requirements of the relevant documented standard now superseded by the term nonconformity.

Nypro Ltd: Site of a major explosion at Flixborough in the United Kingdom in 1973, when a plant modification resulted in an explosion killing 28 people.

occupational health: Health effects that are work related.

occupational illness: Illnesses directly attributable to an individual's exposure to hazards at work.

occupational physician: A qualified medical practitioner skilled in the diagnosis and treatment of occupational illness and occupational health issues.

open question: A question that does not lead to a 'yes' or 'no' answer.

operational audit: Audit examining the compliance with local instructions or procedures.

OSHA: Occupational Safety and Health Administration; the US health and safety regulator.

OHSAS 18001: An international standard for assessing health and safety management performance and the forerunner of ISO 45001.
piezometer: Groundwater monitoring borehole.
Piper Alpha: a major explosion in 1988 on an oil rig in the North Sea which killed 167 oil workers.
Plaudit/Plaudit 2: The Solway Consulting Group's system for audit process management.
Post-it® Notes: Self-adhesive coloured notelets produced by the 3M company.
potable water: Drinking-quality water.
PPE: Personal protective equipment (i.e. gloves, goggles, hardhats, etc.).
product stewardship: The responsible and ethical management of the SHE aspects of a product throughout its life cycle.
pro forma: A standard form or pre-prepared format.
protocol: A pre-prepared checklist used to guide the auditor through the audit discussions and ensure that all key requirements are covered.
quality audit: Systematic and independent examinations to determine whether quality activities and related results comply with planned arrangements and whether these arrangements are implemented effectively and are suitable to achieve objectives.
quality guild: A network of quality-assessed small businesses.
quality manual: Document that defines the quality system in an organisation.
RCRC: The reason–choose–read–challenge method of reviewing documents.
risk: Likelihood that a substance, activity or process will cause harm in the actual circumstances in which it is used.
risk assessment: Process for identifying risks in the workplace, such that risks may be controlled as much as is reasonably practicable.
safe system of work: A formal procedure that results from systematic examination of a task in order to identify all the hazards. It defines safe methods to ensure that hazards are eliminated or risks minimised.
safety inspection: Safety assessment where the assessor uses his or her own knowledge and experience as the criteria for compliance. The inspection usually uses primarily observation skills.
serial audiometry: Hearing surveillance.
scope: The intention or the requirements to be considered.
SHE: Safety, health and environmental.
SHEEMS: Emergency management system for small and medium sized enterprises.
shoring: Provision of supports to prevent the collapse of an excavation.
SPA: Safety performance assessment; a safety assessment focused on a particular topic.
specialist audit: Audits examining a single topic in great depth (e.g. environmental audit or electrical safety audit).
superfund: Colloquial reference to 'CERCLA' (see CERCLA).
stakeholders: People who may either affect or be affected by aspects of SHE management. They may include not only employees but also visitors, contractors, customers and neighbours.
standard: Written requirement that can serve as the basis for comparison.

STOP: Behavioural programme developed by E.I. DuPont de Nemours.
substance: A chemical compound and its impurities, which may be either naturally occurring or manmade.
TECs: Training and enterprise companies.
TWA: Time-weighted average; noise or chemical exposure averaged over an 8-hour period (usually).
underlying cause: Fundamental systemic reason why something happens.
user-friendly: Easy to use; idiot-proof.
validate: To ratify or confirm.
validation: Confirmation of competence.
VDU: Visual display unit (e.g. computer screen).
verification: Process of confirming that things are done the way that people say they are.
wellness: An active process of becoming aware of and making choices towards a more successful existence.
Wiel's disease: An infection carried by rat's urine which infects sewer water.
worker compensation: Term for financial compensation following a work-related injury or illness.
work permit: A formal system that assesses risk and identifies safe controls to allow work to proceed.
workplace: Location where the paid work activities are carried out (e.g. office, workshop, school, home, car).
world-class performance: Sort of health, safety and environmental performance achieved by 'best in class'.
WRULD: Work-related upper-limb disorder.

Appendix 1: Auditor Guidance

This appendix is provided for those wishing to carry out their own audits and provides a quick reference section for essential audit information.

APPENDIX A1.1: SHE ASPECTS FOR CONSIDERATION IN THE AUDIT SCOPE

Audit Subjects

1. Organisation and arrangements for SHE policies
2. Occupational health and hygiene arrangements
3. Management of SHE improvement
4. SHE communication processes
5. Communication of material hazards
6. Control of public statements on SHE matters
7. Training arrangements
8. Chemical inventories
9. Hazard identification and assessment
10. Control of exposure to noise and substances
11. Provision of SHE information to customers

12. Control of biological hazards (*Legionella*, Wiel's disease, etc.)
13. Control of SHE on capital projects
14. Control of modifications and temporary repairs
15. Fire management
16. Provision and maintenance of plant technical information
17. Epidemiology arrangements
18. Safe operation of pressurised systems
19. Lifting equipment
20. The safety of buildings and structures
21. Safety assurance of trips and alarms
22. Safe systems of work arrangements
23. Isolation of plant and equipment from process materials or sources of energy
24. Permits to work and risk assessment
25. Entry into confined spaces
26. Excavation or break-in to walls/ceilings
27. Control of hot work (welding and grinding)
28. Control of sources or ignition in hazardous areas
29. Working on or adjacent to live electrical conductors
30. Control of visitors
31. Lone working
32. Working with asbestos
33. Safe working on roofs
34. Travel and driver safety
35. Manual handling and loading arrangements
36. The use of personal protective equipment
37. Guarding of machines
38. Safe operation of overhead and mobile cranes
39. Safe operation of forklift trucks
40. Abrasive wheels
41. Gas detectors
42. Housekeeping
43. Employee safety awareness campaigns
44. Scaffolding and temporary access arrangements
45. Selection and monitoring of external warehouse
46. SHE arrangements in laboratories
47. Working with visual display terminals
48. Emergency plans
49. Use of contracted services
50. Toll manufacturing
51. Environmental impact assessments
52. Management of effluent and wastes
53. Protection of ground and groundwater
54. Product safety arrangements
55. Arrangements for SHE information reporting

56. Accident investigation
57. Solid waste disposal
58. Control of air emissions
59. Drainage
60. Soak-aways and ditches
61. Landfills
62. Storage tank secondary containment
63. Drum storage
64. Loading and unloading of liquids
65. Groundwater abstraction
66. Ground contamination – Historical review
67. Site investigations
68. Waste minimisation
69. Energy conservation
70. Water conservation
71. SHE auditing arrangements

Note: Subject interviews should be clustered together so that all the subjects relevant to one individual are dealt with together to avoid unnecessary disruption for your managers.

APPENDIX A1.2

Auditor Selection Criteria

All members of the audit team should have

- Formal auditing training
- Prior auditing experience
- Experience of similar activity to that carried out in the audited unit
- A thorough understanding of the relevant regulatory requirements
- Excellent interpersonal skills
- Sufficient seniority to stand up to the local senior manager
- Knowledge of the local language and culture (if overseas)
- Professional SHE knowledge

The lead auditor should also have

- Wide experience of SHE auditing
- Credibility with the audit team
- Credibility with the auditees
- An understanding of the efficient running of the audit process
- Good organisational skills

APPENDIX A1.3

Audit Preparations

The lead auditor is responsible for

- Agreeing on the audit dates
- Agreeing on the scope
- Ensuring that there are auditable standards against which compliance can be assessed
- Agreeing on the audit programme with the audit manager
- Identifying the pre-audit documentation requirements
- Providing suitable audit checklists or protocols
- Compiling the audit manual (if required)
- Chairing the entry meeting
- Managing the audit process
- Optimising the skills and knowledge of the other auditors
- Keeping the auditees informed of progress during the audit
- Chairing the exit meeting
- Compiling and editing the audit report
- Obtaining a copy of the most recent audit report covering the same location and scope

APPENDIX A1.4

Example of Audit Notification Letter

Dear __________

Occupational Health, Safety and Environmental Management Audit

Thank you for the opportunity to carry out a Level __ safety, health and environmental management audit at your site. I would suggest that the audit should be scheduled for the week commencing ________________.

I would propose that the audit programme should follow our usual process of

1. Audit discussions with nominated persons responsible for managing various aspects of health and safety performance
2. Physical condition inspections of most site areas
3. Verification discussions on the plant with staff at all levels to confirm the information gathered during audit interviews

Based on my previous experience, I would suggest that a programme for the audit would be along the following lines:

1. *Proposed audit programme*

   Day 1

   Auditor health and safety induction

Audit entry meeting: All members of the senior management team and any other interested parties should attend this. The entry meeting will be short (20 minutes) and will explain the process that is to be followed. I suggest a starting time of _________.

Management discussions: (You will be asked to nominate one person to talk about each of the topics on the audit scope list.) The nominated people should be the most knowledgeable person on-site on each topic and are not necessarily managers. I shall be interested in hearing about what procedures or instructions exist, what training has been done and how you ensure compliance. I normally allow an average of 15 minutes for each subject.

Day 2

Site tour and verification discussions (starting at _______). I normally start the day following up the topics discussed with managers the previous day. The day will be spent either on the plant or looking at training records, and so on.

Day 3

Management discussions (starting at _______).

Day 4

Site tour and verification discussions: Follow-up of Day 3 management discussions.

Day 5

Site tour and verification discussions (starting at ________). Follow-up of management discussions.

Exit meeting preparation (____ hrs). This is for me to prepare for the exit meeting.

Exit meeting (______ hrs). To be attended by those who attended the entry meeting on Day 1. This meeting usually takes about an hour.

The scope of the audits will include environmental management as well as health and safety, in order to meet your requirements.

2. *Reporting*: Following our normal practice, I usually produce a report similar to the example attached which includes detailed auditor comments as well as the 'Key Recommendations'. However, it is cheaper and sometimes clearer to have a report that is 'Key Recommendations' only. Please let me know your preference. The report is usually submitted initially as a draft to ensure that the audit sampling approach has not picked up something that is non-representative.
3. *Audit scope*: Attached to this letter is an initial checklist of those safety, health and environmental aspects that you may want to include in the scope of the audit. (Final scope to be agreed on between you and me at least 3 weeks before the audit date.)

Please give me a call to let me know if these provisional dates are suitable and so that we can agree on what aspects of your safety, health and environmental management systems you wish to include in the audit scope.

I look forward to having this opportunity to learn from the good practices that you have in place.

Yours sincerely,

APPENDIX A1.5

Typical Sequence of Audit Processes

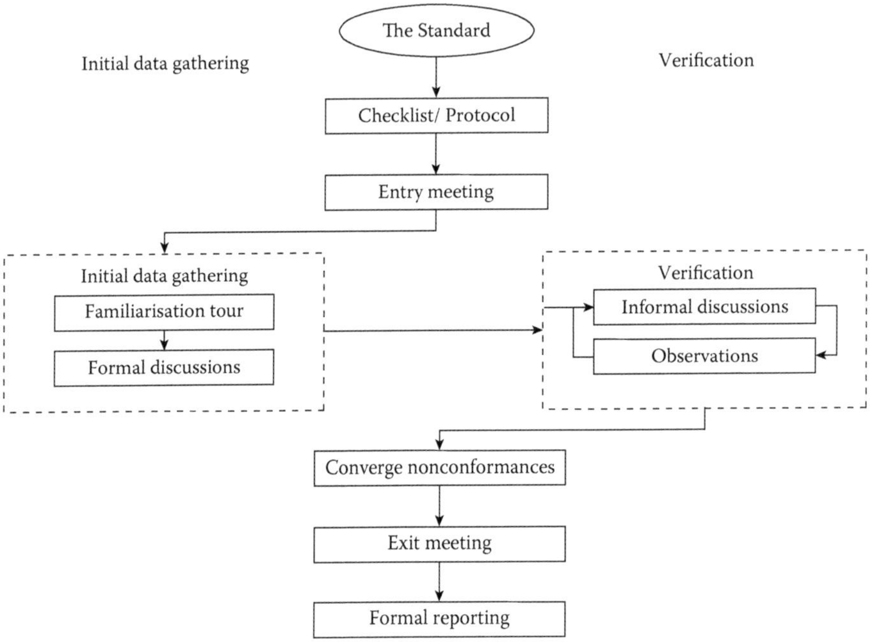

APPENDIX A1.6

Contents of Auditor's Manual

The auditor's manual is a loose-leaf file in which information is compiled for the benefit of the auditors. Typically, it might contain copies of the following:

- Audit notification letter and communications with the auditee
- Audit scope
- Entry meeting presentational material or notes
- Audit programme
- Location layout plan (for large and complex offices or factories)
- Organisation chart of the management of the audited unit
- Previous audit reports
- Auditor's guidance notes or rules
- Checklists or protocols

- Quantitative reporting process if required
- Blank copies of auditor's working papers
- Target numbers of discussions to be carried out

APPENDIX A1.7

Auditor's Personal Equipment

- Appropriate personal protective equipment for the site visits and inspections
- Relevant audit checklist or protocol
- Notepaper
- Pens/pencils
- Highlighter marker pens
- Self-adhesive notes
- Clipboard (to allow note-taking during site visits)

APPENDIX A1.8

The audit checklist should be derived from the standard being audited, by asking the questions identified in the checklist preparation flowchart.

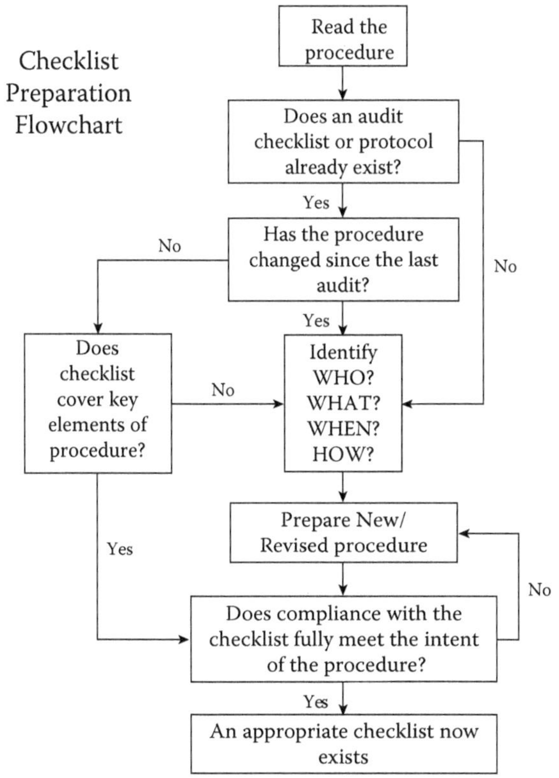

APPENDIX A1.9

Entry Meeting

Purpose:

- Introduce the members of the audit team to the auditee's senior management
- Review the scope and the objectives of the audit
- Provide a short summary of the methods, procedures and programme to be used to conduct the audit
- Establish official communication links between the audit team and the auditee
- Confirm that the resources and facilities needed by the audit team are available
- Confirm the time and date for the closing meeting
- Clarify any unclear details on either side

Possible presentational preparation could require slides to be prepared to indicate

- Purpose of the audit
- Names and background of auditors
- Audit scope
- Audit programme
- Logistics and arrangements
- Reporting arrangements and exit meeting

APPENDIX A1.10

Generic Audit Questions

1. Why does the standard or instruction exist?
2. What legislation applies and is it understood?
3. Are written procedures/instructions in place to ensure that the requirement is met?
4. When were the procedures last reviewed and updated?
5. Are all the responsibilities for managing and carrying out the procedures assigned? (Are the named people still alive?)
6. Are the people who are required to act on this requirement trained and validated?
7. Does local monitoring or auditing regularly assess the degree of compliance with this requirement?
8. What are the latest corrective actions and what is their state of implementation?
9. What are the consequences of failure to comply with this requirement?

APPENDIX A1.11

Discussion Preparation

Attitude and Approach

- Be calm, objective and detached.
- Be courteous, alert and responsive.
- Be friendly and nonaggressive.
- Don't jump to conclusions or make assumptions.
- Don't be judgemental.

Setting

- Go to the auditee's work area.
- Make sure that you are on equal ground.
- Try to keep it one to one.
- Minimise distractions.

Nonverbal Communications

- Shake hands.
- Maintain eye contact.
- Don't invade their 'space'.
- Tolerate silence.
- Check your own understanding.
- Stand beside rather than across a desk.

Types of Question

- Avoid yes/no questions.
- Avoid leading questions.
- Ask open-ended questions.
- Remember the all-important
  - 'How do you know?'
  - 'Show me.'

APPENDIX A1.12

Discussion Questions for Informal Discussions

- Why are you doing that?
- When did management last discuss SHE with you and what did you talk about?
- What training have you had to ensure that you understand the risks of your job?
- Why shouldn't that *liquid be spilled on the floor*?
- How do you make sure that you cannot be harmed by this task?
- Show me where I can find copies of the health and safety instructions.
- Show me how you would isolate that equipment.

- Show me what you would do if *the fire alarm sounds.*
- Show me what protective equipment you use to do this job.
- What are the risks associated with this job?
- How have risks of injury/environmental harm associated with this task been controlled?

Note: Words in italics are given as examples only.

APPENDIX A1.13

Audit Observations

Remember audit observations entail focused looking, which is derived from physical evidence.

- Observe both behaviours and conditions.
- Notice the unusual.

Observations are usually reliable evidence, but remember the limitations of

- Abnormal behaviour/conditions
- Time of audit (Are you seeing a representative sample?)

Look in out-of-the-way places:

- At (i.e. directly at the item being studied)
- Above
- Beyond
- Behind
- Beneath

Remember that our observations are influenced by

- Our experiences
- Our training
- Our interests

Finally, do you understand what you are seeing?

- Is it an optical illusion?
- Does it need more explanation?
- Is it illogical?
- Does it comply with the standard?

APPENDIX A1.14

Reporting

Check in what format the auditee requires the report to be compiled.

- Pro forma style (most likely for Level 1 audits)
- Conventional free text report (most likely for Level 2 and 3 audits)

Before compiling the report:

- Be sure of your facts.
    - Base recommendations on evidence.
- Provide a balanced response.
    - Positive recognition.
    - Opportunities for improvement.
- Separate 'noncompliances' from 'observations'.
- Recognise the scale of your recommendations.
- In reviewing your findings, see if there are any significant trends or patterns; these will be more important than individual discrepancies.
- Try to avoid making issues out of trivial errors.

Avoid:

- Generalising
- Vagueness
- Legal opinions
- Using unfamiliar terminology
- Criticism of individuals

Finally, a free text report might have the following sections:

1. Acknowledgements
    a. Appreciation for help and cooperation provided during the audit
    b. Lead auditor's appreciation of work done by audit team
2. Introduction
    a. Name and role in the organisation of the location being audited
    b. Indication of the scale/size of the operation being audited
    c. Date and duration of audit
    d. Names and job titles of auditors
    e. Date of last audit
3. Scope of audit
    a. Statement of the purpose of the audit
    b. List of the SHE aspects covered by the audit
    c. Indication of who agreed on the scope

4. Executive summary
    a. Short summary of main outcomes of the audit, including
        i. Areas of excellence
        ii. Opportunities for improvement
5. Main recommendations
6. Detailed findings (optional; include this only at the specific request of the auditee, as this will significantly extend the length of the report)
    a. A brief sentence or paragraph relating to the auditor's findings for each aspect
7. Scoring (only if mandated by the organisation)
8. Arrangements for follow-up of actions

Keep it simple.

APPENDIX A1.15

Example of Executive Summary for a Level 3 SHE Management Audit Report

Executive Summary

This audit of the safety, health and environmental management systems at ______________ Ltd _________ site was carried out on request by ________________.

Local management and employees are to be congratulated for the high level of environmental awareness observed within the company, for the quality of healthcare systems and for the excellent fire safety management processes. Further areas of excellence were observed in the areas of engineering, where design and equipment standards are high and maintenance information procedures and routine SHE assurance work were found to exceed the requirements. Detailed work in the areas of *Legionella* monitoring and control, laboratory procedures and VDU assessments are all worthy of specific recognition.

The desire to improve was clearly in evidence, as seen by recent work on new control procedures for contractors and the eagerness with which some new ideas were immediately acted upon during the audit.

The primary concern of the auditor is that the systematic approach, which is used to control production quality, is not extended into the SHE area. Although some formal procedures do exist, they are not consistently communicated and enforced. This is particularly concerning in the 'safe systems of work' area, where isolation standards are not well observed.

There is a need to establish robust safety, health and environmental systems that are enforced through a routine programme of local auditing.

Management accountability for SHE is not totally clear. The practice of focusing safety matters, in particular, through the safety representatives and safety manager appears to bypass line management. Every opportunity should be taken to reinforce the concept that SHE management is a line management responsibility that is supported by functional professionals and representatives. Greater communication of SHE matters including 'learning points' could go a long way towards raising SHE awareness at all levels.

Training standards were found to be patchy and threatened by recent organisational changes. It is important that training 'needs' are clearly identified for all individuals, such that skills and knowledge shortcomings may be recognised and supplemented. The importance of maintaining effective sustainable training records must be recognised and fully implemented.

Although engineering SHE assurance is generally of a high standard, arrangements should be put in place to ensure the routine proof testing of critical safety and environmental protection trips. The modification control and project risk analysis systems are in need of strengthening.

Personnel are well aware of 'food contact' controls for product health and safety, but there is little knowledge of other product regulatory matters. Although the products are of a relatively benign nature, evidence should be available to assure customers that all relevant regulatory requirements have been met.

Housekeeping standards within the factory were observed to be in need of attention. This problem should be addressed urgently to avoid injuries arising. Housekeeping is an ideal area for management to demonstrate commitment in a way that will involve all employees.

APPENDIX A1.16

Process Safety Audit Aspects

The following aspects should be covered in any full process safety audit.

1. Management commitment
2. Process safety information
3. Process hazard analysis and risk assessment
4. Operating procedures
5. Safe systems of work
6. Employee training and competence assurance
7. Management of contractors
8. Pre-startup safety review
9. Asset integrity
10. Non-routine work authorisations
11. Managing change
12. Incident investigation
13. Emergency preparedness
14. Compliance-level auditing

APPENDIX A1.17

Pre-Startup Safety Checks

Ensure that

- All equipment identified on the drawings is present and installed correctly.
- All protective systems been correctly installed (e.g. instrumented trips, relief valves).

- All trips and alarms have been tested.
- All previous process hazard study actions been either resolved or completed (with no new hazards introduced).
- All plant operating instructions are available for all modes of operation including normal, abnormal and emergency conditions.
- All personnel have received appropriate training.
- All equipment and procedures necessary to protect the environment and for monitoring environmental performance are in place.
- All plant-based safety equipment is in place (e.g. showers and eyewash stations).
- There is adequate access for operations and maintenance.
- Firefighting equipment, such as hoses and extinguishers, is in place.

APPENDIX A1.18

EHS Due Diligence: Areas for the Auditor to Review

1. The EHS policy of the organisation
2. The EHS management system
3. A hazard register or equivalent
4. EHS procedures and instructions
5. List of hazardous materials in use
6. Description of the risk management system
7. Compliance with licences and authorisations
8. Records of injuries and incidents
9. EHS responsibilities/organisation chart
10. Reports to regulatory bodies
11. Employee handbook
12. Safety training records
13. Ground and groundwater investigation reports
14. Site histories
15. Environmental reports
16. Waste disposal arrangements
17. Current and potential litigation
18. List of insurance claims
19. Emergency management plan
20. Fire management arrangements

    If the organisation has significant process-related risks (see Chapter 30) then the following should be added to the list:
21. Process flow diagrams
22. Plant dossiers
23. Process hazard assessment records
24. Records of the periodic testing of safety critical systems
25. Records of the control of design and other changes
26. Maintenance records and evidence of periodic integrity testing of critical equipment

This list should be considered to be a basic starting point for the EHS due diligence auditor.

APPENDIX A1.19

Due Diligence: Environmental Audit Checklist

1. Current use of the property
2. Historic usage of the property
3. Current and historic usage of neighbouring properties
4. Hydrogeological map of the area
5. Geological conditions
6. Descriptions of structures
7. Roads and railways
8. Water supply: piped potable and groundwater
9. Drainage systems (foul and chemical)
10. Hazardous substances list
11. Biological substances
12. Underground storage tanks, sumps and pits
13. Aboveground storage
14. Evidence of spillages
15. Drums storage areas
16. Odours
17. Vegetation damage
18. Septic tanks and effluent treatment systems
19. Ponds and lagoons
20. Wells and boreholes

Appendix 2: Plaudit 2 Audit Protocol

Aspect No.	
001	Arrangements for safety, health and environmental policies
002	Occupational health and hygiene arrangements
003	Management of SHE improvement
004	SHE communication processes
005	Communication of material hazards
006	Control of public statements on SHE matters
007	Training arrangements
008	Chemical inventories
009	Hazard identification and assessment
010	Control of exposure to noise
011	Control of exposure to respiratory hazards
012	Provision of SHE information to customers
013	Control of biological hazards
014	Control of SHE on capital projects
015	Control of modifications and temporary repairs
016	Fire management
017	Provision and maintenance of technical information
018	Safe operation of pressurised systems
019	Safety of buildings and structures
020	Trips and alarms
021	Safe systems of work arrangements
022	Isolation of equipment from process materials or sources of energy
023	Permits to work and risk assessment
024	Entry into confined spaces
025	Excavation or break-in to walls/ceilings
026	Control of hot work (welding and grinding)
027	Control of sources or ignition in hazardous areas
028	Control of visitors
029	Lone working
030	Working with asbestos
031	Safe working on roofs
032	Travel and driver safety
033	Manual handling and loading arrangements
034	Use of personal protective equipment
035	Guarding of machines
036	Safe operation of overhead and mobile cranes
037	Safe operation of forklift trucks

038 Abrasive wheels
039 Housekeeping
040 Employee safety awareness campaigns
041 Scaffolding and temporary access arrangements
042 Selection and monitoring of external warehouse
043 SHE arrangements in laboratories
044 Working with visual display terminals
045 Emergency plans
046 Use of contracted services
047 Product safety arrangements
048 Environmental impact assessments
049 Solid waste disposal
050 Control of air emissions
051 Drainage
052 Soak-aways and ditches
053 Landfills
054 Storage tank secondary containment
055 Drum storage
056 Loading and unloading of liquids
057 Groundwater abstraction
058 Ground contamination: historical review
059 Site investigations
060 Waste minimisation

ASPECT 001 SHE Policy

001	Audit Check	How to Verify	Act	Notes	OK
-01	Is there a current SHE policy statement?	View statement.			
-02	Does the statement cover SHE?	Check for specific references to environmental, safety and occupational health management.			
-03	Is the statement signed by the current senior manager?	Identify the most senior manager (CEO?) and ensure the statement carries his or her signature.			
-04	Does the statement include details of key SHE responsibilities?	Check that the named people are still here and that those mentioned are aware of their responsibilities and are acting on them. Check for missing names.			
-05	Has the policy been communicated to all associates?	Check that the policy is on display in places where associates will see it.			
-06	Are associates aware of the policy?	Ask a random sample of employees if they are aware of the policy, where it can be found and what it says.			
-07	Is the policy subject to periodic review?	Confirm that the policy is a 'living' document and that it is reviewed regularly (within the last 3 years) or after significant changes or learning events.			
-08	Does the policy cover all aspects of the operation?	Does it include the SHE effects of the total business – for example, product safety, disposal, travel safety, community outreach, etc.?			
-09	Is the policy complied with?	Check health and safety statistics and environmental measures (emissions/waste etc.) to confirm if the facility is serious about implementing the policy.			
-10	Is the policy in clearly understood language?	Ask a random selection of associates what the policy means to them and how it applies to their job.			
				Aspect 001 total	

Aspect 002 Occupational Health and Hygiene Arrangements

002	Audit Check	How to Verify	Act	Notes	OK
-01	Are persons appointed with responsibility for SHE management?	Talk to the named individuals. Are they fully aware of their responsibilities?			
-02	Are the responsibilities of the appointed person clearly written down?	Does the appointed person know and carry out these responsibilities?			
-03	Is the appointed person suitably trained and experienced?	Check training records. Does the person have an appropriate external qualification? Is he or she suitably knowledgeable and experienced in the activities carried out at this facility?			
-04	Does the appointed person carry appropriate authority?	Is there evidence that the management team responds to issues raised by the appointed persons? Do they have the ear of the most senior manager?			
-05	Does the facility have an occupational physician appointed?	Check if the physician actually visits the facility on a regular basis (at least weekly) and whether the appointee has occupational health qualifications.			
-06	Is there a medical room on-site for minor medical treatments?	Visit the room and check if the facilities are suitable for the anticipated medical emergencies.			
-07	Does the facility have a first-aider (ERT) always in attendance?	Check who is the first-aider on duty today and what qualifications he or she has.			
-08	Is personal protective equipment provided to control residual hazards to health?	Visit protective equipment store. Is appropriate equipment available to all users? Check if associates are wearing equipment in designated areas.			
-09	Is health and hygiene information communicated at all?	Is there a communication cascade process? Do associates have an SHE consultation process? Ask associates when they last attended a safety meeting.			
-10	Do people know and understand the relevant SHE law?	Check how the facility stays up to date with changes in the law. Check if regulatory audits are carried out.			
				Aspect 002 total	

Aspect 003 Management of SHE Improvement

003	Audit Check	How to Verify	Act	Notes	OK
-01	Does a structured process exist to manage improvements in SHE?	Talk to managers at the highest level to assess their commitment and personal involvement in managing SHE improvement.			
-02	Do improvement targets exist?	View the targets and ask where the current performance is against the target.			
-03	Are injury records maintained and used for accident prevention?	Examine the records. Are they the legally reportable injuries only, or are all minor injuries also recorded and used for prevention?			
-04	Does an occupational hygiene programme exist?	Check if the hygiene monitoring plans are being fully carried out. Are these plans adequate for the nature of activities at the facility? Are safe limits ever exceeded?			
-05	Is there a waste minimisation plan in place?	Check what wastes are produced and what recycling plans are in place. Ask to see evidence of the last few years' waste performance.			
-06	Is there a need to monitor atmospheric emissions?	Check if legal limits are applied to the emissions. Check the last 2 months' emissions-monitoring records and ask about noncompliances.			
-07	Do regulatory controls apply to wastewater discharges?	Check if the discharges are to a sewer or to water courses. How are rogue discharges prevented? Check monitoring results and any noncompliances.			
-08	Does a medium-term (3–5 years) SHE improvement plan exist?	View the plan. Check for progress against outstanding actions.			
-09	Are progress reviews carried out periodically?	Check if the management team routinely reviews SHE performance and improvement plans and assess whether this results in action.			
-10	Are associates involved in planning SHE improvement?	Check action plans to see if this is just management led. Talk to associates about how they see themselves involved.			
				Aspect 003 total	

Aspect 004 SHE Communication Processes

004	Audit Check	How to Verify	Act	Notes	OK
-01	Do arrangements exist for SHE communication?	Check to see if there is a structured cascade process which regularly cascades SHE information from management to all associates and stakeholders.			
-02	Is information cascaded in an effective way?	Talk to associates. Did they remember and understand the key points from the last communication?			
-03	Is all the information available to associates necessary for them to carry out their jobs?	Talk to a spectrum of different associates. Ask if they have access to sufficient information to do their jobs. Where would they go to get additional information? Can they read and understand it?			
-04	Is communication two-way?	Is there a suggestions scheme? Ask associates for SHE examples of where managers listen or don't listen to what they say.			
-05	Are the minutes of SHE meetings available to all?	Check noticeboards etc. for examples of minutes being displayed. Talk to associates about whether minutes are read.			
-06	Is the learning from all injuries passed on to all?	Talk to associates. What was the last learning event that they were told about? What have they learned?			
-07	Does a process exist for associates to flag SHE concerns?	Check what the process is and whether it has been used recently. Do associates know it exists?			
-08	Are records maintained to ensure key communications reach everyone?	Check communication records to ensure that important messages are communicated to those absent on the day of the initial communication.			
-09	Are changes in instructions and procedures communicated?	Does a process exist for communicating changes to essential operational and SHE instructions, including the reason for the change?			
-10	Is SHE information communicated to neighbours?	Check how it is done. Talk to a sample of neighbours. Is the organisation perceived as a good neighbour? Are there any current issues?			
				Aspect 004 total	

ASPECT 005 Material Hazards

005	Audit Check	How to Verify	Act	Notes	OK
-01	Is there an inventory of hazardous materials in use at the facility?	During the site inspection, observe chemicals in use and ask to see the materials safety data sheet (MSDS) and check if it is on the inventory.			
-02	Does an MSDS exist for all chemicals in use?	Who maintains the MSDS files? Check if they are up to date. Do associates understand them and know where to access them?			
-03	Have the risks of using hazardous materials been assessed?	Ask to see examples of assessment for chemicals that you select from the inventory. Are the defined controls in place?			
-04	Are material hazards effectively controlled?	Select a number of commonly used and less commonly used chemicals and check if they are being handled in accordance with the MSDS.			
-05	Do associates understand the hazards of materials?	Talk to associates and ask if they know the hazards of the materials that they are using and how those hazards should be controlled.			
-06	Are materials correctly stored?	Examine a selection of both bulk and small-quantity chemicals to check if they are stored in a way likely to prevent injury or incident.			
-07	Are engineering controls inspected and maintained?	Where engineering controls are required to control exposure (e.g. fume cupboards, local extraction, etc.) check records to ensure that these are regularly inspected and tested.			
-08	Do associates need periodic medical or exposure monitoring?	Check if any chemicals in use are subject to periodic medical surveillance or if dosimeter checks are required. Check if these are actually happening.			
-09	Do associates understand exposure limits?	Talk to associates and test their understanding of workplace exposure limits and the effects of time-weighted averaging.			
-10	Is there a process to control the introduction of new substances?	Check with purchasing people if arrangements are in place to prevent the purchase or sampling of new substances until an MSDS is available and risk assessed.			
				Aspect 005 total	

ASPECT 006 Control of Public Statement on SHE Matters					
006	**Audit Check**	**How to Verify**	**Act**	**Notes**	**OK**
-01	Is there an approved spokesperson to talk to the media on SHE matters?	Who is it? Is there at least one deputy? Talk to them.			
-02	Was the spokesperson selected on the basis of the impact he or she has?	Has the spokesperson been selected on the basis of public empathy or just seniority? Does he or she have sufficient SHE competence to talk knowledgeably?			
-03	Is the spokesperson trained in media relations?	Examine training record or course attendance certificate. What practical experience has the person had since training?			
-04	Is there a chosen media interview site on the facility?	Has thought been given as to where to hold interviews and why? (The chief execs' office may not carry the right message.)			
-05	Are key staff trained in what action to take if the press makes contact?	Is there a policy on whether to admit press or answer phone calls from the media? Check if key administrative assistants/receptionists/security are aware.			
-06	Does the facility make a point of issuing regular good news?	Check media files to see copies of latest press releases and whether they led to media coverage.			
-07	Is good SHE performance immediately obvious?	Observe the facility from the point of view of the passer-by. Is it obvious that good SHE performance is important (signs, visitor induction)?			
-08	Are plans in place to deal with the media consequences of an emergency?	Check to see if a pre-prepared media release exists and whether this goes to the media, local dignitaries, emergency responders, head office and employees.			
-09	Are plans in place to respond to unsolicited approaches from the media?	Check plans. Is there any responsibility allocated to keep track of local environmental activists or any company activities that may lead to local opposition?			
-10	Is there a risk to the public from defective products?	Check how the organisation would respond to public concern about products. Is there a product recall and damage limitation plan?			
				Aspect 006 total	

ASPECT 007 Training Arrangements

007	Audit Check	How to Verify	Act	Notes	OK
-01	Are training needs clearly identified for individuals?	Check if a training needs analysis has been done for all associates.			
-02	Does a training programme exist for all associates?	Select associates at random and ask to see their training plan or programme. It's a good idea to ask to sample these at all levels (check that the boss has one).			
-03	Is SHE induction training carried out for all new associates?	Check records. Look particularly for records for contract, agency and casual workers.			
-04	Are training records for all associates maintained?	View records. Select individual names at random and explore the detail of their training. Check if records are up to date.			
-05	Are records of training specifications and content maintained?	Check records to see if it is clear exactly what each trainee has been trained in (i.e. is the training suitable and sufficient?).			
-06	Is all training subject to competent validation?	Check if training is formally validated and check competence/ qualification of validators.			
-07	Are trainees closely supervised before they are fully competent?	Talk to associates who are still undergoing training and assess their level of supervision/mentoring.			
-08	Are associates trained to an externally verifiable standard?	Check qualification certificates/training records where external standards apply.			
-09	Is refresher training carried out where necessary?	Check records associated with those activities (e.g. forklift truck driving) where there is a need for routine refresher training.			
-10	Does a process exist to ensure training is provided when things change?	Check whether training relates to the current situation. Equipment may have changed or procedures may have been revised.			
				Aspect 007 total	

ASPECT 008 Chemical Inventories

008	Audit Check	How to Verify	Act	Notes	OK
-01	Does an inventory exist of all chemicals used or stored on the site?	During site tours, note a sample of what chemicals are in use. Pay particular attention to the small quantities used in workshops or samples in offices.			
-02	Does a procedure exist to prevent unauthorised chemicals being used?	View the procedure. Has this been applied in the case of those chemicals you identified during your site inspection?			
-03	Does the inventory record approximate quantities and location?	Do the quantities roughly agree with your observations during the site tour?			
-04	Is there an MSDS available for all chemicals used?	Check who maintains the inventory and keeps the records of MSDS. Is this information readily available to the users?			
-05	Are chemicals stored in a suitable way?	Check standards of bulk storage installations and also minor chemical stores in such places as laboratories, workshops and cleaner's cupboards.			
-06	Are chemical hazards clearly identified at the storage point and container?	During site inspection, check hazard identification labels. Do users recognise the hazards, and are proper controls in place?			
-07	Is the chemical inventory available to the emergency services?	Check if the inventory is available to the fire service in the event of an emergency on-site and, if so, how often it is updated.			
-08	Do associates understand the hazards of what they handle?	Check that associates understand the hazards and risks associated with the hazardous chemicals that they are using. Are the controls adequate?			
-09	Is transportation of chemical samples in private cars forbidden?	Check if procedure or policy exists to prevent the transportation of chemical samples in private cars. If it exists, how is it enforced?			
-10	Does the chemical inventory include intermediates?	If the facility manufactures or blends chemicals, does the inventory include those chemicals which are partially produced or blended?			
				Aspect 008 total	

ASPECT 009 Hazard Identification and Assessment

009	Audit Check	How to Verify	Act	Notes	OK
-01	Is there a list of the key hazards associated with this facility?	Check if the list is thorough, selecting some hazard you have seen at random. Does the list include business travel or hazards from adjacent facilities?			
-02	Have all hazards been identified and risks assessed?	Ask to see some risk assessments relating to work in progress. Are the assessments out-of-date generic ones or are they really specific to the task in hand?			
-03	Have hazards been identified systematically?	Check for evidence of hazard studies, job safety analysis, noise assessments or other systematic ways of hazard identification.			
-04	Are hazard controls in place?	Check for use of the hierarchy of controls: (1) elimination, (2) substitution, (3) engineering controls, (4) system of work, (5) training, (6) personal protective equipment (PPE).			
-05	Is management responsible for managing hazards?	Check if hazard management is management led or whether associates are left to identify all the hazards for themselves.			
-06	Is suitable PPE available to control residual hazards?	Check the quality and availability of PPE. Is it suitable and are associates trained in its use and limitations?			
-07	Does hazard identification include environmental hazards?	Check for evidence that routine hazard and risk assessment includes environmental risk.			
-08	Is hazard and risk identification done by qualified persons?	Check if a list of qualified hazard and risk assessors exists and whether their training and validation is appropriate.			
-09	Are risk assessments reviewed when circumstances change?	Check if reviews are carried out if the associate changes (e.g. from experienced to inexperienced or from male to female – that is, different strength or physique).			
-10	Do situations exist where the task stops altogether because the hazard is too great?	Ask when the last time it was that a task did not proceed at all because assessment viewed it as too hazardous. Does it ever happen? Or are those jobs left until night shift?			
				Aspect 009 total	

ASPECT 010 Control of Exposure to Noise

010	Audit Check	How to Verify	Act	Notes	OK
-01	Are there places at the facility where noise is a problem?	Ask what the noise level is. Compare this with the legal permitted levels.			
-02	Are noise surveys carried out?	View the results of recent surveys. Do they record what was going on at the time – that is, was all or part of the installed equipment running at the time?			
-03	Are noise contour maps available?	View the contour maps. Do they cover all the known noisy areas?			
-04	Are all noisy work areas clearly identified on-site?	Check signage/designations on-site and that these agree with the noise maps.			
-05	Do associates always wear suitable hearing protection?	During site inspection observe compliance of people in noisy areas with the hearing protection rules. (Remember to wear your own hearing protection.)			
-06	Do workers in noisy areas have serial audiometry tests?	Check if audiometric tests are available and what action is taken by management if noise-induced hearing loss is detected.			
-07	Is effective emergency communication present in noisy areas?	Ask how associates would hear an emergency evacuation alarm. Look for visual alarms to supplement the usual audible alarm system.			
-08	Is noise dosimetry used for transient workers?	Where associates are exposed to variable noise levels, check if there is a need for personal noise sampling (dosimeters).			
-09	Do associates understand noise limits, dBA and time-weighted averages (TWA)?	Ask associates in noisy areas if they have been trained in understanding noise limits, the dBA scale and TWA.			
-10	Do associates receive a pre-employment hearing test?	Check if a hearing baseline is set when associates join the organisation. Otherwise, the company may be liable for hearing loss predating this employment.			
				Aspect 010 total	

ASPECT 011 Control of Exposure to Respiratory Hazards

011	Audit Check	How to Verify	Act	Notes	OK
-01	Is exposure to dust and fumes controlled?	What exposures exist and how are they controlled?			
-02	Are all exposure hazard areas clearly identified on-site?	Check signage/designations on-site.			
-03	Do associates always wear suitable respiratory protection?	Observe compliance with respiratory protection in the workplace. Is fit testing done?			
-04	Do workers at risk of chemical exposure need routine medical checks?	Is a programme of health checks under way for the associates who could be affected? How are those associates identified?			
-05	Are hygiene monitoring checks done routinely to monitor exposure levels?	Check what monitoring is carried out and whether it is area or personal monitoring. Check that the people carrying out the sampling and analysis are properly qualified.			
-06	Do associates understand exposure limits TWA?	Ask associates in exposed areas if they have been trained in understanding exposure limits, TWA and what limit apply to them.			
-07	Are personal breathing apparatus (BA) sets used on the site?	Check why this is. Is it for emergency use (e.g. a site fire service) or is it routinely used in the course of daily operations?			
-08	Are BA sets maintained by a competent person?	Who does the checks? Are they competent? How is usage controlled?			
-09	Are BA users clearly authorised and trained?	Check the authorisation list and training records. Do users go through mask fit training and checks?			
-10	Are BA users subject to frequent medical check-ups?	View the programme of check-ups. Has anyone missed an appointment but still using a BA set?			
				Aspect 011 total	

ASPECT 012 Provision of SHE Information to Customers

012	Audit Check	How to Verify	Act	Notes	OK
-01	What SHE information needs to be provided to customers?				
-02	Do products require a declaration of safety?	Check if an assessment has been done to ensure that the product is safe. In Europe, do products carry a CE mark and declaration of conformity?			
-03	For chemical products is an MSDS available?	View the MSDS.			
-04	Is packaging suitable for the product?	Does the packaging carry a hazard phrase and safety phrase in the language of the country in which it will be used?			
-05	Does the packaging carry safe handling information?	Examine a sample of product packages.			
-06	Does the packaging carry recycling information?	Examine a sample of product packages.			
-07	Does bulk packaging carry ergonomic information?	Check for indications of weight, lifting points, stacking limits and 'way up' markings.			
-08	Are product support contact details available?	Is there evidence of a helpline (telephone number or email address)?			
-09	Is transport emergency advice available?	Check if this is available in all areas where the product is distributed. Is it needed and available outside office hours?			
-10	Is there a procedure for the safe handling and return of empties?	Check if the empty container returns process is functioning effectively.			
				Aspect 012 total	

ASPECT 013 Control of Biological Hazards

013	Audit Check	How to Verify	Act	Notes	OK
-01	Are there arrangements in place to control biological hazards?	Check the procedure exists. In particular, explore such things as Wiel's disease risk from working in or adjacent to drains or sewers.			
-02	Do evaporative cooling towers pose a biological threat?	Check for evidence of effective algae control treatment in the cooling tower ponds.			
-03	Are any products or research arrangements biologically hazardous?	Check if effective controls are in place. Can extraction systems transfer the hazard elsewhere?			
-04	Is there an effective vermin control scheme in place?	Check how often the bait is laid and carcasses are removed. Ensure that bait does not cause a hazard to associates or the general public.			
-05	Are first-aiders (ERTs) or medical staff exposed to blood-borne pathogens?	Check that there is an effective means of protecting medical staff. Are 'sharps' embargoed? Similar protection will be required for public service workers who may be exposed to dirty needles.			
-06	Is there an inoculation programme to control biological health risks?	Check if associates and medical staff are aware of what inoculations are available and what their limitations and side effects are. Are they up to date?			
-07	Are arrangements in place to deal with bites/stings?	Check if there is a known risk of bites or stings from vermin or wild animals and what antidote/treatment would be required.			
-08	Is there a private potable water supply?	Check that there are effective means of purification and that its effectiveness is proved by routine samples.			
-09	Do biological hazards arise from effluent/sewage treatment?	Check for evidence that treatment process is functioning correctly.			
-10	Is kitchen/mess room waste disposed of in sealed containers?	Check kitchen and food preparation areas for signs of poor waste disposal standards that could lead to vermin or unhygienic/biological hazards.			
				Aspect 013 total	

ASPECT 014 Control of SHE on Capital Projects

014	Audit Check	How to Verify	Act	Notes	OK
-01	Are all capital projects subject to a hazard study?	Check hazard study records. Are all relevant stages of the hazard study (studies 1–6) completed? Check if all actions are completed.			
-02	Do projects have a list of regulations pertinent to the project?	View a sample of the lists.			
-03	Has an environmental impact assessment been carried out?	Check if assessments are always carried out. Are these required for local authority approval or environmental permits?			
-04	Has a list of additional health and safety reviews been produced?	Check if list is thorough. Consider noise, substances, manual handling, safe maintenance, hygiene, relief and blowdown, temporary works, etc.			
-05	Has the project considered the SHE aspects of start-up?	Ask to see examples of commissioning plans. Has adequate time been allocated for this activity? Does commissioning include personnel training?			
-06	Are any statutory notifications required?	Check if the appropriate regulatory bodies have been informed about the project or its construction.			
-07	Have new operating instructions been provided?	Check if the project has allowed for the inclusion of new instructions and the training of associates in those instructions.			
-08	Is a full dossier maintained of the technical information?	Ask to see examples of project technical data files. Where the project is carried out by a contractor, ask to see its project manuals.			
-09	Are SHE requirements a part of the purchasing specifications?	Check if purchases specify SHE requirements. For example, do motors have a maximum noise spec or an energy efficiency spec?			
-10	Is the SHE standard specified for equipment handover?	Check contracts to see if SHE performance limits have been set and what evidence exists to show that these were met.			
				Aspect 014 total	

ASPECT 015 Control of Modifications and Temporary Repairs

015	Audit Check	How to Verify	Act	Notes	OK
-01	Is there a procedure to ensure that changes are properly controlled?	Examine the procedure. Does the definition of a change include people changes as well as technical changes? Is there an authorisation process?			
-02	Is the procedure being applied to all modifications?	Check a random selection of completed modifications. Are the assessments thoroughly completed and installed as designed?			
-03	Are modifications subject to suitable authorisation?	Check if modification designers and approvers are competent in the relevant aspects of the design (i.e. qualified engineers of the right discipline).			
-04	Is there a pre-commissioning check before start-up?	Do a random check to see if the modification was installed as designed and that this was checked before start-up. Were operators trained?			
-05	Are changes included in the plant data records and drawings?	Check a selection of master drawings to ensure that the changes are incorporated into the master technical records.			
-06	Are temporary repairs considered modifications?	Check if such things as temporary clamps on leaking pipework or temporary construction works are included in the modification procedure.			
-07	Are there arrangements to deal with emergency modifications?	Does the procedure allow for a shortcut in the process in an emergency? Check if this is still done by competent people and retrospectively recorded.			
-08	Do all changes carry a unique identification?	Select a modification at random from the records and then try to find it during site inspection tours.			
-09	Are changes carried out to the same standard as the original equipment?	Look for evidence of the original design specifications or equivalent being used in the modification.			
-10	Is there a regulatory check before start-up?	Does a competent person consider whether the change meets all the relevant regulatory requirements? Is this person different from the designer; that is, is there a double check?			
				Aspect 015 total	

ASPECT 016 Fire Management

016	Audit Check	How to Verify	Act	Notes	OK
-01	Has a fire risk assessment been carried out for the premises?	Ask to see a copy of the fire risk assessment. Was this carried out by a competent person? Have the recommendations been fully implemented?			
-02	Is there a fire alarm system that can be acted upon by all?	Ask to be present at a test. Can the alarm be heard everywhere? Particularly examine noisy and remote areas. Are visual warnings also required?			
-03	Is the alarm routinely tested?	Check for records of weekly alarm tests. Ask when the last full evacuation test was done (annually for all?).			
-04	Is there a fire evacuation procedure?	View the procedure. Do associates and visitors know what to do in the event of a fire? Ask if they know the location of their evacuation assembly point.			
-05	Is there any unusual fire risk associated with the premises?	Ask if there are any flammable substances in use either in the premises or in the neighbourhood. What precautions are taken? Are they suitable?			
-06	Is suitable firefighting equipment available?	Is equipment well maintained and subject to periodic inspection? Check inspection dates on firefighting equipment during plant inspection tours.			
-07	Are automatic fire suppression systems regularly checked?	See sprinkler test records. Do these include test operation of clack valves? Ask how smoke alarms and rate-of-rise detectors are tested.			
-08	Are escape routes well maintained and identified?	Check during plant tour to see if fire doors will open and whether escape routes are well signposted. Does emergency lighting work/get tested?			
-09	Is there a means of accounting for all in an emergency?	Ask to be shown how the facility accounts for all occupants in an emergency – including transients such as visitors, contractors and the public.			
-10	Does the external fire service visit the premises?	Check for a fire certificate or equivalent. Does it reflect building changes? Is the fire service included in fire management exercises?			
				Aspect 016 total	

ASPECT 017 Provision and Maintenance of Technical Information

017	Audit Check	How to Verify	Act	Notes	OK
-01	Do up-to-date records exist of equipment and building design?	See if records are maintained and whether there is evidence that they are used.			
-02	Do records include maintenance histories and SHE events?	Check if any SHE critical equipment regularly leads to accidents, breaking down or underperforming and whether this leads to any improvement action.			
-03	Is the technical information available to those who need it?	During on-site discussions with associates, check whether they are aware of the technical information and how often they use it.			
-04	Is the technical information adequate?	Does it include such things as design drawings, manufacturers operating and maintenance manuals, SHE assessments, modification records, permits, etc.?			
-05	If the facility is leased, is there access to technical information?	Check who has responsibility for what. Are these clearly identified? Do these include responsibilities for SHE liabilities and SHE information?			
-06	Does SHE information have to be provided with products?	Are product technical files required? Do products require verification (FDA approval), certification (i.e. electrical) or marking (e.g. marking)?			
-07	Do products provide appropriate user information?	Check product packaging or manuals for evidence of user instructions and risk and safety phrases. Is this user-friendly?			
-08	Is product technical help available?	Try calling the helpline. Is the responder able to answer obvious SHE questions?			
-09	Is product failure information maintained?	Check what is available. Is this of a standard to use for anticipated failure rates for customer hazard analysis assessments?			
-10	Is SHE data made available to the community?	Is environmental emission and noise data available for the community? Is this in a form which is easily understandable?			
				Aspect 017 total	

ASPECT 018 Safe Operation of Pressurised Systems

018	Audit Check	How to Verify	Act	Notes	OK
-01	Does a procedure exist to control the design and use of pressure systems?	Check the existence of the procedure. Does it relate to both pressure vessels and pressurised pipe work systems? Is it up to date?			
-02	Does a register exist of all pressurised systems?	Identify pressurised systems during the site tour and check that these are on the inventory. Ask if recent new equipment has been included.			
-03	Are all pressure systems protected by a pressure relief device?	Are safety valves in evidence on pressure systems? Are their set pressures equal to or below the safe working pressures of the pressure system?			
-04	Does a register exist of all pressurised relief devices?	Identify devices during the site tour and check that these are on the inventory. Does the inventory include atmospheric system protection (swan necks etc.)?			
-05	Do all pressure systems have a thorough periodic inspection?	Check that there is a scheduling system to plan the inspections and that none is overdue. Are the inspections carried out by a competent person?			
-06	Do all pressure relief devices have a periodic inspection and retest?	Check that there is a scheduling system to plan the inspections and tests and that none is overdue. Are the inspections carried out by a competent person?			
-07	Is a competent person appointed for design verification?	Ask who is responsible for approving pressure system designs and modifications. What are their qualifications?			
-08	Are arrangements in place for change of duty of pressure equipment?	Check if second-hand equipment is ever used or if equipment is ever used for purposes other than those it was designed for. Ask how safety is ensured			
-09	Does inspection include inspection of the system peripherals?	On piping systems are the pipe supports/hangers, insulation or trace heating checked? On vessels, do checks include sights glasses, vibration dampers, etc.?			
-10	Is there a robust process to remedy actions identified at inspection?	Check how the inspection recommendation is acted upon. Take a recent inspection report and follow through to see if the work is completed.			
				Aspect 018 total	

ASPECT 019 Safety of Buildings and Structures

019	Audit Check	How to Verify	Act	Notes	OK
-01	Is responsibility for building maintenance clearly defined?	Check if building is leased or owned. Check who has responsibility for what. Are these clearly identified? Do responsibilities include upkeep and alteration?			
-02	Are regulatory requirements being met?	Who controls planning and building control applications to the authorities? Are the statutory welfare requirements being met?			
-03	Is there a procedure to control changes to the building structure?	Check that these are authorised by a competent person and that no changes are occurring without approval.			
-04	Is a competent person appointed for design verification?	Ask who is responsible for approving structural changes. What are their qualifications? (This is probably an external architect; talk to the architect and see what he or she has done recently at the site).			
-05	Do all buildings and structures have a periodic inspection?	Check that there is a scheduling system to plan the inspections and that none is overdue. Does a competent person carry out the inspections?			
-06	Are walls etc. scanned for services before drilling/cutting into them?	Does the procedure require a scan for buried pipes or electric cables before drilling or cutting into walls and surfaces? Check that the scanner works.			
-07	Are safe floor loadings clearly marked?	Check for evidence of floor loading during site tour. Ask how associates know whether they are exceeding the floor loading.			
-08	Is there a procedure for testing for asbestos in buildings?	Is there a record of where asbestos exists within the building? Are the locations clearly marked?			
-09	Is there a procedure to control the installation of temporary buildings?	Check if portakabins and trailers have been through any planning or safety assessment.			
-10	Is the building secure when unoccupied?	Check what the security arrangements are. Is the building locked and patrolled when unoccupied? Who monitors fire alarms etc.?			
				Aspect 019 total	

ASPECT 020 Trips and Alarms

020	Audit Check	How to Verify	Act	Notes	OK
-01	Is there a register of all SHE critical trips and alarms?	View the list.			
-02	Are trips and alarms regularly tested?	Examine the procedure or policy which calls for periodic testing.			
-03	Is there a formal process for setting test frequencies?	Ask the person who manages the trip and alarm test programme how test frequencies are decided. Does it seem logical, suitable and sufficient?			
-04	Is there a schedule for the testing of trips and alarms?	Examine the schedule. Are any tests overdue? Explore the reasons for tests going overdue. Is it a one-off reason or endemic?			
-05	Are tests fully functional?	Check if tests are just circuit tests or whether they instigate the intended function (i.e. when tested, does the trip actually stop the motor?).			
-06	Is testing carried out by a competent person?	Check whether the testers are properly trained, competent and authorised.			
-07	Is there a process for correcting faulty trips and alarms immediately?	Check records to see if faults are permanently remedied.			
-08	Is there a process for tracking repeated failures?	Check whether repeat faults have been a short-term fix or whether they address the long-term underlying problem.			
-09	Does the alarm checking cover vehicles and mobile plant?	Ask to be shown how this is recorded.			
-10	Are lamps on enunciators regularly checked?	Is there a system of lamp-checking trip and alarm enunciator lights on a regular basis (once/day/shift) to check for bulb failures?			
				Aspect 020 total	

ASPECT 021 Safe Systems of Work Arrangements

021	Audit Check	How to Verify	Act	Notes	OK
-01	Is there a procedure defining the safe systems of work?	View the procedure. Does it cover all aspects of the enterprise, including office e-work and travel?			
-02	Does the system include the provision of suitable operational procedures?	Select a task at random and check that it has an associated instruction which refers to the SHE requirements to complete the task.			
-03	Does the system cover non-routine tasks (e.g. maintenance/contract)?	Check if a 'permit to work' process exists and whether it covers all maintenance, contractor and non-routine tasks.			
-04	Does the safe system specify the need for effective isolations?	Is there an effective 'lock-out' or other system to ensure that energy and process fluids are effectively isolated and risks minimised?			
-05	Is there a safe system to ensure the safety of visitors?	Check that visitors (this may include yourselves) to the site are properly inducted and aware of the main hazards and emergency arrangements.			
-06	Does associates' behaviour confirm that they work safely?	During the tour, observe associates' behaviour. Are they wearing the appropriate personal protective clothing?			
-07	Does the safe system include the provision of safe plant and equipment?	During the tour, observe equipment standards. Are they well maintained and suitable for their purpose?			
-08	Is there a safe system of work for business travel?	Check for evidence of business travel safety risk assessments. Are there daily road driving limits and are the lowest-risk modes of travel employed?			
-09	Are hazards identified and risk controlled?	Look for examples of a systematic approach to controlling hazards and managing risk.			
-10	Are the safe systems regularly audited?	Ask to see recent audit reports and the audit plan. Follow up a sample report to see if actions are completed.			
				Aspect 021 total	

ASPECT 022 Isolation of Plant and Equipment from Materials or Energy

022	Audit Check	How to Verify	Act	Notes	OK
-01	Does a policy exist relating to the need for adequate isolation?	View the policy or procedure. Does it cover both process fluid isolation and energy isolation standards?			
-02	Do competent persons carry out isolations?	Is there a list of approved persons? Have they been trained and validated?			
-03	Is it forbidden to work under isolations performed by programmable systems?	Check that competent persons always carry out isolations and that these cannot be overridden by automation systems such as computers or other forms of automatic control.			
-04	Are isolations tamper proof?	Are isolations locked off with the key held only by the authorised person or person engaged in the work (i.e. some form of lock-out system)?			
-05	Is there an energy isolation plan for each task?	Look at a job in progress. Ask to see examples of the forward thinking that went into the isolations associated with the task.			
-06	Is the effectiveness of the isolations tested before starting work?	Does the procedure require a lock, tag and try concept (i.e. after isolations are done is the start button tried to see if the drive starts)?			
-07	Is there a formal process for multiple isolations?	Check a task that involves either multiple energy isolations or both fluid and energy isolations to see if these are coordinated and cross-referenced?			
-08	Is there a process to ensure that work is complete before isolations are removed?	Ask to see what checks are made to ensure that the task is completed before energy or fluids are reinstated.			
-09	Are all points of isolation equipped for lock-off?	Check a random sample during the tour. Office machines may rely on physical disconnection.			
-10	Are all energy and fluid sources clearly identified?	Check this during the plant tour.			
				Aspect 022 total	

ASPECT 023 Permits to Work and Risk Assessments

023	Audit Check	How to Verify	Act	Notes	OK
-01	Is there a formal permit system to control work?	Check if this applies to all tasks other than those for which a formal instruction exists or where there is a formal exemption.			
-02	Are issuers of permits trained and authorised?	Check training records. See if non-authorised persons ever issue permits to work.			
-03	Do the permits clearly specify the task to be done?	Check a random sample of the permits.			
-04	Does the permit identify all the hazards and assess the risks?	See if hazards are always clearly identified on the permit form.			
-05	Does the permit identify isolations required to do the work?	Check if isolations are clearly specified and carried out by a competent person.			
-06	Does the permit identify how to control residual hazards?	Check permit for details of how hazards are to be controlled. Look particularly for protective equipment requirements. Are they always adopted?			
-07	Is there a formal authorisation step in the process?	Is the permit always authorised by a trained 'issuer' and accepted by a trained 'acceptor'?			
-08	Is there a clearly defined time limit for each permit?	Examine a permit to see if there is a clearly stated expiration date and time. Is this in line with the procedure?			
-09	Are users of the permit familiar with the stated requirements?	Check some tasks in progress and ask the workers what are the permit requirements.			
-10	Is the permit process regularly audited to ensure compliance?	Ask to see examples of recent audit reports. Have the corrective actions been implemented?			
				Aspect 023 total	

ASPECT 024 Entry into Confined Spaces

024	Audit Check	How to Verify	Act	Notes	OK
-01	Is there a clear formal definition of a confined space at this location?	Check that the definition covers all locations where there is a risk of dust and fumes causing a hazard.			
-02	Is there a formal permit to authorise entry into a confined space?	View the permit instruction and examples of completed permits.			
-03	Is a rescue plan always available?	Ask to be shown examples of recent rescue plans. Have these ever been tested through a rescue exercise?			
-04	Are suitable air tests always carried out?	Ensure that the air test always tests for oxygen but also for other anticipated gases like flammables, CO, H_2S, etc.			
-05	Are gas analysers regularly calibrated?	See the recent calibration record. Does a competent person do the calibration?			
-06	Has the space been isolated from hazard sources?	Are piped services physically disconnected and have sources of energy, heat and cold been removed?			
-07	Does the task introduce hazards into the space?	Check if the nature of the task being performed introduces hazards that render the space unsafe (i.e. inert welding gases etc.).			
-08	Are entries permitted using breathing apparatus?	Check if there are procedures in place to ensure that BA sets are properly maintained, and users are trained, competent and have mask fit tests.			
-09	Is adequate air circulation provided?	What are the arrangements for clean air circulation in the space? Are air movers of 'suction' rather than 'blower' design?			
-10	Are confined spaces identified and have access controlled?	Check for the presence of signs. Is there a physical gate or guard to prevent access when an entry permit is not in place?			
				Aspect 024 total	

ASPECT 025 Excavations or Break-Ins to Walls and Ceilings

025	Audit Check	How to Verify	Act	Notes	OK
-01	Are adequate checks carried out to identify buried services?	Check procedure. Does it call for drawing/records check and inductive (CAT) scans to identify the presence of live conductors and metal pipes?			
-02	Are arrangements in place to prevent falls into excavations?	Examine an excavation. Is it adequately fenced? Is safe access provided into the excavation?			
-03	Is a permit required to commence an excavation or break-in?	View a sample of permits.			
-04	Are permit issuers and users trained and competent?	Are permit issuers trained in recognising and controlling the hazards associated with excavations and break-ins? Are users also trained?			
-05	Is the excavation shored to prevent collapse?	Observe the shoring standards on an excavation.			
-06	Are checks carried out to ensure the absence of risks from ground contaminants?	What sorts of checks are carried out? Are users aware of the risks from ground contaminants and biodegradation (hydrogen sulphide)?			
-07	Is the excavation treated as a confined space?	What precautions are taken? Do users carry personal gas monitors?			
-08	Are risks from mobile plant and engine exhausts controlled?	Can exhaust gases affect workers in the excavation? Is the mobile plant operating on stable land?			
-09	Are arrangements in place to control the presence of water?	Examine the risks from drowning, slipping, mobile pump exhaust gas and disease.			
-10	Do excavation controls apply to floors, walls and ceilings?	Does the procedure consider the weakening effect and premature collapse of part or all of the structure?			
				Aspect 025 total	

ASPECT 026 Control of Hot Work (e.g. Welding, Grinding, etc.)

026	Audit Check	How to Verify	Act	Notes	OK
-01	Is there a clear formal definition a 'hot work' at this location?	Check that the definition covers welding, burning, grinding and all other applications of naked flames.			
-02	Is there a formal permit to authorise hot work?	View the permit instruction and examples of completed permits.			
-03	Is there a competent issuing authority?	Check the training and experience of the person who issues hot work permits. Is there a trained and experienced deputy?			
-04	Does the procedure require the presence of a standby person?	Check if the standby person understands his or her responsibilities. Does the person always have a fire extinguisher available?			
-05	Is the work area checked for combustibles?	Check a hot work job in progress. Has the area been adequately checked to exclude the presence of combustibles?			
-06	Do workers wear the correct personal protective equipment?	Are welders using welding screens, gauntlets and protective aprons? Do grinders wear gauntlets and protective aprons and dark goggles?			
-07	Is the air tested for the presence of flammable gases?	Ask to see the flammable gas detectors. Are they recently calibrated and in a suitable position?			
-08	Are passers-by adequately protected?	Check if screens or tenting are erected to prevent the effects of welding 'flash' or grinding sparks. Is access below the job barriered off?			
-09	Are worker sources of ignition controlled?	Ask how burners ignite their equipment. Do they use spark guns?			
-10	Are welders aware of gas and fume hazards?	Talk to welders and ask them if they understand the hazards of welding flux fume or inert gases (argon or CO_2).			
				Aspect 026 total	

ASPECT 027 Control of Sources of Ignition in Hazardous Areas

027	Audit Check	How to Verify	Act	Notes	OK
-01	Are area classifications carried out?	Ask to see the drawings that define the electrical classification of each area.			
-02	Is the area classification up to date?	Ask about a recent equipment change in a hazardous area. Does this change the classification? Has this been incorporated in the area classification drawing?			
-03	Is electrical equipment in hazardous areas subject to inspection?	Ask to see the inspection schedule. Are all the inspections up to date?			
-04	Do procedures exist to control the use of temporary equipment?	What controls are required to introduce temporary or portable electrical equipment into hazardous areas? Are hot work permits issued?			
-05	Do purchasing specs define area classification requirements?	Discuss orders for equipment in hazardous areas with the purchasing clerks. Do they understand its relevance?			
-06	Is electrical earthing (grounding) adequate?	Are earth straps required across flanged joints? Are all earthing straps in place as required by the equipment design specs?			
-07	Is static electricity controlled?	Are bulk containers grounded during offloading? Do associates wear antistatic footwear? Do they understand static generation from steam leaks etc.?			
-08	Is there a de-matching policy?	Is there a means of controlling the use of personal items like matches, cigarette lighters, mobile phones, calculators, etc. in the hazardous area?			
-09	Are flammable gas vents fitted with flame arrestors?	Check the existence of some flame arrestors. Are they routinely inspected to ensure that they don't become blocked?			
-10	Are lightning conductors effective?	Do lightning conductors exist and are they checked for continuity?			
				Aspect 027 total	

ASPECT 028 Control of Visitors

028	Audit Check	How to Verify	Act	Notes	OK
-01	Is there a process to ensure the safety of visitors?	Do visitors sign in and sign out? What about regulars like delivery drivers and postmen? Did everyone sign out yesterday?			
-02	Are visitors always accompanied by a company representative?	Are you as an auditor a visitor? Are you always accompanied or have you been given appropriate training?			
-03	Do visitors know what to do in an emergency?	Ask a visitor.			
-04	Are visitors given an appropriate safety induction?	Is there any system for checking that they have understood the basics?			
-05	Are visitors aware of the primary hazards at this location?	Ask a visitor. Do they understand both safety and occupational health hazards?			
-06	Is protective clothing provided for visitors where appropriate?	Do visitors wear protective equipment in the appropriate places?			
-07	Do all visitors have a prearranged appointment?	Would it be possible for someone to gain access as a visitor with malicious intent?			
-08	Are visitors accounted for in an emergency?	Are visitors told where their assembly point is and to whom they should report?			
-09	Is the health and safety culture clear to visitors?	Is the safety culture visible? Look for notices and visible publicity such as accident statistics, safety committee minutes, posters, warning signs, etc.			
-10	Are visitors clearly identifiable?	Do visitors wear badges or clearly labelled coveralls or hats?			
				Aspect 028 total	

ASPECT 029 Lone Working

029	Audit Check	How to Verify	Act	Notes	OK
-01	Is 'lone working' a recognised issue at the location?	Do any workers operate alone in remote locations where there could be difficulty in communication if anything went wrong?			
-02	Is there a procedure for lone worker emergency response?	Is it possible for associates to have a medical or safety emergency at work without management being aware in a reasonable time?			
-03	Is there a regular check on all associates during the working period?	Do supervisors/managers check on the well-being of all their staff during the working day or shift?			
-04	Do lone drivers know what action to take in the event of emergency?	How would management know if there had been an accident or medical emergency while drivers are away from their base? Does anyone check if they are OK?			
-05	Is there regular contact with lone travellers?	Ask how the health and safety of travellers is ensured.			
-06	Is there an assessment of associates at high risk when working alone?	Are associates with certain medical conditions (e.g. epilepsy, heart conditions, blackouts, etc.) required to always work in pairs or have some form of immobilisation alert?			
-07	Is there an assessment of experience before associates work alone?	Do individuals undergo a competence validation before being permitted to work alone?			
-08	Is lone working taken into account in task risk assessments?	View the risk assessments. Contact a lone worker and ask if he or she has valid risk assessments with them.			
-09	Do service contractors have lone working safety checks?	Particularly vulnerable groups are out-of-hours cleaning contractors and security personnel.			
-10	Is the lone worker alarm system periodically tested?	This may be a proprietary lone worker alarm or a telephone/beeper system. Ask what the response time was at the last test.			
				Aspect 029 total	

ASPECT 030 Working with Asbestos

030	Audit Check	How to Verify	Act	Notes	OK
-01	Is there a register of asbestos-containing materials at the location?	View the register. Does it contain records of asbestos insulation, ceiling tiles, fire-resistant panels and blankets, gland packing, asbestos cement sheets, etc. containing asbestos?			
-02	Are asbestos-containing materials clearly marked?	During site inspection note the presence or absence of asbestos labels.			
-03	Are asbestos materials properly sealed?	Is asbestos insulation outer sealant in good condition to prevent the escape of free fibres?			
-04	Is there a procedure for the removal of asbestos?	Is removal limited to licensed contractors? Check the contractor's licence.			
-05	Is there a clear procedure for the disposal of asbestos?	Is asbestos waste sealed in clearly marked impermeable bags? Are bags sealed before removal from the enclosure? Is waste sent without delay to a licensed disposal site?			
-06	Is asbestos removal only done within a sealed enclosure?	Is the enclosure checked and approved before use to stop fibre leakage? Is a filtered air supply provided? Is it connected directly to a changing room/shower?			
-07	Are air tests carried out before the enclosure is removed?	Ask to see examples of past fibre counts. Does an approved laboratory do the test analysis?			
-08	Are asbestos workers subject to regular medical surveillance?	Talk to asbestos workers. Ask when they had their last health check-up.			
-09	Do asbestos workers have suitable welfare arrangements?	Do asbestos workers have separate changing facilities for clean and dirty work wear? Is there a hot water shower facility between them? Is contaminated work wear disposed of to waste?			
-10	Do asbestos workers wear suitable respiratory protection?	Check respirators are suitable for asbestos.			
				Aspect 030 total	

ASPECT 031 Safe Working on Roofs

031	Audit Check	How to Verify	Act	Notes	OK
-01	Is access to roofs formally controlled?	Check for access control notices or procedure.			
-02	Does the roof access procedure consider the risk of falls?	Check procedure.			
-03	Are the edges of roofs suitably barriered when work is in progress?	Observe a selection of flat roofs.			
-04	Is safe, well-secured access provided to the roof?	Check for presence of stairs/ladders.			
-05	Are crawling boards always used to access fragile roofs?	Where are the crawling boards? Ask to see them.			
-06	Are arrangements made to prevent falls through skylights?	Ask how skylights are protected. Talk to workers who work on roofs to see what controls they actually use.			
-07	Do roof workers always use suitable fall arrest equipment?	Do workers understand the limitations of fall arrest equipment? How often is the equipment checked? What are used as anchors?			
-08	Are all fragile roofs clearly labelled?	Look for labels during site tour.			
-09	Does the procedure apply to working in loft spaces and above ceilings?	The hazards of falling through ceilings are similar to those of fragile roofs. Check if there is full body access to any lofts or above ceilings.			
-10	Have relevant workers been warned of the hazards of roof working?	Talk to roof workers and ask what sorts of accidents happen on roofs. Have they been told of the frequency of fatalities in this type of work?			
				Aspect 031 total	

ASPECT 032 Travel and Driver Safety

032	Audit Check	How to Verify	Act	Notes	OK
-01	Does a policy exist to address the health and safety aspects of travel?	View the policy.			
-02	Is there a clear limit on how far associates may drive in a day?	Check travel and mileage claim forms. Are business travellers complying with the policy?			
-03	Is the frequency and duration of rest periods defined for travellers?	Check tachograph records for heavy-goods drivers. Do car travel departure and appointment times allow for rest periods?			
-04	Are drivers' driving licences checked before travel is authorised?	Is there a process to ensure that drivers have the appropriate type of vehicle licence? Are there periodic checks to ensure that the licence is still valid?			
-05	Are company vehicles well maintained?	Is there a scheduled maintenance programme for vehicles, and is it up to date?			
-06	Are associates required to use the lowest-risk mode of transport?	Are risk assessments carried out for business travel? Are alternatives such as audio conferences etc. used whenever possible?			
-07	Are appropriate immunisations provided for overseas travel?	Is there a requirement for medical advice before travelling overseas? Are immunisations made available? Talk to travellers to see what advice they receive.			
-08	Is there a formal process for dealing with travel emergencies?	Ask what would happen if an associate had a medical emergency while overseas. Is there a method of providing independent medical advice?			
-09	Do travellers understand the health risks of travel?	Talk to travellers. Do they know the risks of deep vein thrombosis, tiredness, drinking water, dietary change, sexually transmitted disease, etc.?			
-10	Are associates trained in hotel evacuation safety?	Talk to travellers. Do they check exit routes? What should they do in the event of a fire outside their room?			
				Aspect 032 total	

ASPECT 033 Manual Handling Arrangements

033	Audit Check	How to Verify	Act	Notes	OK
-01	Are manual handling operations that involve risk of injury avoided?	Is there a policy or procedure in place?			
-02	Are manual handling operations properly assessed?	View some of the risk assessments. Has a trained and competent person carried these out?			
-03	Are associates informed of the results of manual handling assessments?	Check how associates are made aware of the workplace controls necessary to reduce the risks associated with manual handling hazards.			
-04	Are associates trained in kinetic lifting techniques?	Observe associates doing lifting. Are they applying the principles of kinetic lifting (e.g. bending knees, straight back when lifting items)?			
-05	Are the weights of items that are lifted clearly identified?	Ask associates how they know the weights of items that they lift. Examine production areas, warehousing, goods inward and maintenance areas.			
-06	Is installed lifting/handling equipment correctly used?	During the site inspection observe how and if associates make full and proper use of the installed handling equipment.			
-07	Are associates aware of their own manual handling limitations?	Talk to associates. Do they understand how stature, position and repetition can effectively reduce the weight of what can be safely handled?			
-08	Do product designs consider the handling requirements?	Check product safety files.			
-09	Are manual handling risks considered in goods purchasing?	Check if this is incorporated into purchasing specifications or whether it is just done on an ad hoc basis.			
-10	Are manual handling risks in offices considered?	Check with office staff.			
				Aspect 033 total	

ASPECT 034 Use of Personal Protective Equipment (PPE)

034	Audit Check	How to Verify	Act	Notes	OK
-01	Is the need for PPE based on risk assessment?	See examples of risk assessments.			
-02	Is PPE considered to be a last line of defence?	Ask workers why they wear protective equipment and how decisions are made about what PPE is required.			
-03	Is PPE a 'personal' issue?	Check if equipment is issued on a personal rather than shared basis. If equipment is shared, what are the hygiene arrangements?			
-04	Are associates fit checked for their PPE?	Are hardhat sweatbands correctly adjusted? Are respirators checked for good contact/fit? Is equipment too large/too small?			
-05	Are associates trained in the safe removal of contaminated PPE?	Ask individuals how they would remove a pair of contaminated goggles without causing injury.			
-06	Are records maintained of personal issues of PPE?	Ask to see the records. Are they updated when there are new technical developments in PPE quality or standards?			
-07	Is PPE inspected and well maintained?	Is there a procedure to ensure that all PPE is maintaining in a condition that is fit for purpose?			
-08	Are PPE purchases specified by a competent person?	Check if a safety specialist or other competent person specifies PPE.			
-09	Is the wearing of PPE routinely monitored?	Check with managers about how they ensure that associates wear the specified protective equipment.			
-10	Are PPE requirements clearly labelled in the work areas?	Check if signs are visible, clearly understood and follow normal conventions.			
				Aspect 034 total	

ASPECT 035 Guarding of Machines

035	Audit Check	How to Verify	Act	Notes	OK
-01	Is there a register of all physical guards at the location?	Check the register. Is there a clear method of physically identifying the guard shown in the register?			
-02	Are guards periodically checked?	Does the quality of the check ensure that the integrity of the guard system is maintained as intended in the original design?			
-03	Is there a periodic function test on all guard interlocks?	Check if guard interlocks are tested in a safe manner and at the beginning of each working period.			
-04	Are guards designed to prevent contact with moving parts?	Check a sample of guards. Visually examine (don't touch) whether the guards prevent penetration, over- or under-reaching.			
-05	Are light beam and pressure pads considered as guards?	Are all beams and pad operations tested on a regular basis?			
-06	Is there a check on guard integrity after maintenance?	Is this requirement written into the work control permitting procedure?			
-07	Can routines such as lubrication and speed checks be done without guard removal?	Ask how couplings are lubricated and how speed measurements and rotation checks are done.			
-08	Are guarding standards applied to mobile plant?	Examine mobile plant during site inspections.			
-09	Are guards attached with tamper-proof fixings?	Guards should require a tool (spanner) for removal, in order to reduce the chance of removal by operators. Be vigilant for missing bolts, wing nuts and lever clamps.			
-10	Are electrical guards in place?	Check that electrical cabinet doors are closed and junction/fuse box covers are in place.			
				Aspect 035 total	

ASPECT 036 Safe Operation of Cranes

036	Audit Check	How to Verify	Act	Notes	OK
-01	Are all cranes clearly marked with the safe working load?	Don't check only the major cranes. Look for derricks, hoists and runway beams.			
-02	Are cranes on a register or list?	Examine the register.			
-03	Are cranes subject to a periodic thorough examination?	Check if a competent person does the examination. Are any inspections out of date? How do operators know that the crane in use has been examined?			
-04	Is lifting gear on a register and periodically examined?	Check the register. Are all items of lifting gear (i.e. strops, shackles, lifting eyes, chains, pull lifts, etc.) in good condition and examined by a competent person? Are there any overdue examinations?			
-05	Are all crane drivers and users trained and competent?	Check training records.			
-06	Are ropes and chains changed periodically?	Check maintenance records.			
-07	Do procedures exist to check credentials of contract crane drivers?	Pay particular attention to contract mobile cranes when used on-site. Are drivers' training certificate and crane latest test certificates checked?			
-08	Is there a formal process for checking ground stability?	This is for mobile cranes. Are outriggers or wheels on stable ground when lifting? Are load spreading plates used? Has the travel route been checked to ensure it is load bearing (i.e. no weak culvert etc.)?			
-09	When travelling with loads are banks men used?	This normally applies to overhead travelling cranes. Is someone appointed to ensure that the suspended load does not collide with people or equipment?			
-10	Is there a policy of not standing under suspended loads?	Ask to see the policy statement. Do your observations confirm that this is followed?			
				Aspect 036 total	

ASPECT 037 Safe Operation of Forklift Trucks (FLTs)

037	Audit Check	How to Verify	Act	Notes	OK
-01	Are all FLTs clearly marked with the safe working load?	Observe markings on trucks.			
-02	Are trucks fit for purpose?	Check that internal combustion engines are not used in enclosed and unventilated spaces. Are aisles wide enough for the type of truck used?			
-03	Do competent persons periodically inspect trucks?	Check inspection records.			
-04	Are truck masts and lifting components periodically checked?	Check inspection records.			
-05	Are all drivers suitably trained and validated?	Ask a selection of drivers to show their FLT licences.			
-06	Are daily safety checks carried out on all trucks?	Examine daily or shift records.			
-07	Do drivers observe safe driving practices?	Observe operating speeds, use of audible warnings and safe manoeuvring.			
-08	Is recharging or refuelling carried out in a safe place?	On electrical trucks ensure charging bays are well ventilated to disperse hydrogen and are in good condition. Ensure gas and liquid fuels are safely stored and bunded (dyked).			
-09	Are pedestrians segregated from FLT traffic?	Check locations of walkways in FLT areas.			
-10	Are FLT trucks well maintained?	Check maintenance records. Examine trucks for evidence of collision damage.			
				Aspect 037 total	

ASPECT 038 Abrasive Wheels

038	Audit Check	How to Verify	Act	Notes	OK
-01	Are all grinding machines clearly marked with max. speed?	Label should be on the machine or immediately adjacent. Does this speed agree with the speed marked on the wheel or disc?			
-02	Are all grinding machines/grinders on a register or list?	Examine the register. Does it include handheld tools as well as fixed machines?			
-03	Does a competent person install wheels?	Check the training records and experience of the competent person.			
-04	Are grinding machines correctly adjusted and well maintained?	Visually examine some fixed grinding machines. Are the wheels dressed and not subject to eccentric wear? Are the work piece rests close to the wheel? Are guards and visors in place?			
-05	Are spindle speeds checked periodically?	Check speed records. Can tachometers be used without guard removal?			
-06	Are goggles/face visors specified as personal protection?	Look for signs by machine or a safety procedure.			
-07	Are portable grinders subject to periodic inspection and maintenance?	Electrical machines need to have portable appliance testing. Ask to see examples of these tools. Pay particular attention to the condition of flexes and air hoses.			
-08	Are frequent users of grinders checked for evidence of hand–arm vibration (HAV) syndrome?	Ask what arrangements are in place to check for vibration white finger or HAV syndrome (very common in metal fabrication).			
-09	Is protection provided for adjacent workers and passers-by?	Does the grinding procedure specify the need for screens and distance barriers to protect other people?			
-10	Are fire hazards controlled?	Grinding provides a source of ignition. Ask how this fire hazard is controlled.			
				Aspect 038 total	

ASPECT 039 Housekeeping

039	Audit Check	How to Verify	Act	Notes	OK
-01	Is housekeeping maintained to high standards?	Check if there are formal arrangements to ensure housekeeping standards are maintained. Ask what is considered to be an acceptable standard.			
-02	Are housekeeping inspections carried out periodically?	Ask to see the housekeeping inspection forward plan. Is guidance/ training provided for those carrying out the inspection?			
-03	Are housekeeping actions dealt with promptly?	Check the process for completing actions. Check how many outstanding actions remain from the last few inspections.			
-04	Are all personnel made aware of the importance of housekeeping?	Discuss this with associates. Do they understand the link between housekeeping standards and injury potential?			
-05	Do senior managers set a good example of housekeeping?	Are the senior managers' offices housekeeping standards safe and setting a good example? Do they take part in housekeeping inspections?			
-06	Is there a policy of cleaning up after a job is completed?	Examine some recently completed jobs.			
-07	Are storage facilities adequate?	Is there a 'place for everything and everything in its place' or are materials stored in the work area?			
-08	Are walkways clearly identified?	Ensure that walkways and emergency routes are kept clear.			
-09	Are spillages promptly cleaned up?	Check product, lubricant and waste skip/dumpster areas.			
-10	Do facilities exist for the satisfactory disposal of waste materials?	Check if waste bins, trash cans, skips/dumpsters are available and there is a system for waste disposal.			
				Aspect 039 total	

ASPECT 040 Employee Awareness Campaigns

040	Audit Check	How to Verify	Act	Notes	OK
-01	Is people's behaviour recognised as important in incident prevention?	Explore what proportion of injuries/incidents are recognised to be caused by unsafe or environmentally unfriendly behaviour.			
-02	Are safe or unsafe behaviours measured?	See the measurements and look for examples of improving trends.			
-03	Is there an established and formal behavioural improvement process?	Check if the process identifies behaviours or just unsafe/ environmentally hazardous conditions.			
-04	Are there up-to-date safety and environmental statistics displayed?	Check the SHE reporting notice boards. Are they up to date with relevant information for the expected reader?			
-05	Are arrangements in place to highlight known hazards?	Look for examples of posters, photographs and safety or environmental messages on display. Are they up to date and eye-catching?			
-06	Are accident and incident reports available to all?	Ask associates how they learn about the messages from incident investigations.			
-07	Are annual statistics used to plan improvement actions?	Ask to see injury, occupational illness and environmental incident statistics. Find out how any trends are acted upon.			
-08	Are associates involved in defining awareness campaigns?	Check safety/environmental committee minutes. Find out who leads the improvement campaigns.			
-09	Are awareness campaigns based on 'leading' as well as 'lagging' indicators?	Lagging indicators are things like injuries and past events. Leading indicators might be something like striving towards a future achievement of one million hours without an injury.			
-10	Is the effectiveness of awareness campaigns measured?	Ask how the management team knows which campaigns were successful and which were not. Look for evidence that they have learned from this.			
				Aspect 040 total	

ASPECT 041 Scaffolding and Temporary Access Arrangements

041	Audit Check	How to Verify	Act	Notes	OK
-01	Is some means of safe temporary access used whenever working at heights?	Look for evidence of scaffold registers, use of lanyards and mobile elevating working platforms. Standard ladders do not usually constitute a safe workplace unless the work duration is very short and the ladder is secured.			
-02	Do trained and competent personnel only erect scaffolding?	Check a sample of scaffolder training records.			
-03	Are all scaffolds inspected before use?	Check some scaffolds and review the inspection record. Is it still valid?			
-04	Do all temporary access platforms have a safe means of access/exit?	Look for evidence of dedicated ladder access on scaffolds. Is there a method of preventing access until the first inspection is completed?			
-05	Is a check done to ensure that the scaffold is fit for its purpose?	Check the intention of the access. A scaffold can be structurally sound, but may be too low to allow the work to proceed safely.			
-06	Has the required load on the scaffold been assessed?	Check if the scaffold has been designed just for one man working, or will it need to have a load-bearing capability?			
-07	Is the area beneath scaffolds protected against falling objects?	Check if this area is barriered off (or some other appropriate form of personnel protection) to prevent injury from falling objects.			
-08	Do trained personnel only operate mobile working platforms?	Check that procedures allow only trained operators to drive powered mobile, elevating working platforms.			
-09	Are personnel working at height required to wear fall harnesses?	Observe whether workers using lanyards and harnesses are correctly clipped on to an appropriate immovable object.			
-10	Are all ladders and step ladders registered and periodically inspected?	Check a selection of ladders. Are they numbered and do they have a valid inspection?			
				Aspect 041 total	

ASPECT 042 Selection and Monitoring of External Warehouses

042	Audit Check	How to Verify	Act	Notes	OK
-01	Are arrangements in place to assess the SHE implications of external warehousing?	Check the procedure or other formal arrangements to see what aspects of SHE hazards have been considered. Is a manager appointed to monitor these arrangements?			
-02	Is the warehouse suitable for storing this material?	If materials are chemical or biological substances, consider what could happen in the event of loss of containment.			
-03	Could a release of this material affect operators or the public?	Look for evidence that the warehouse operators are suitably trained to handle this material in an emergency.			
-04	Is some form of licence or authority needed to store this material?	Check to see if the warehouse holds a valid authorisation.			
-05	Is storage suitable to prevent impact damage?	Check if the storage area is segregated from general vehicular or forklift truck traffic.			
-06	Is there a suitable means of fire suppression?	Is the fire system just an alarm or will it automatically extinguish the fire? What is the expected turnout time for the fire brigade?			
-07	Could other materials stored in the warehouse adversely react with our materials?	Is there a clear understanding of what else is currently or could be stored in the warehouse? Has the possibility of chemical interactions been risk assessed?			
-08	Is there effective fire segregation in the warehouse?	Check the segregation and that the function of fire doors is periodically tested.			
-09	Are the warehouse management's SHE standards checked?	Ask what aspects of SHE are checked. Are the warehouse management's intentions borne out in practice?			
-10	Is the site notified of all incidents in external warehouses?	Ask to see all the notifications in the last year.			
				Aspect 042 total	

ASPECT 043 SHE Arrangements in Laboratories

043	Audit Check	How to Verify	Act	Notes	OK
-01	Is unauthorised access to the laboratory effectively controlled?	Check if laboratory doors are kept locked when unoccupied and arrangements are in place to restrict entry to authorised persons.			
-02	Are harmful substances kept under control?	Check how oxidising agents, carcinogens, radioactive sources are stored. Are quantities at the minimum required (or don't exceed the maximum permitted by law)?			
-03	Are accurate records maintained of all experimentation?	View records and laboratory notebooks.			
-04	Have all substances in use been assessed to determine their hazard to workers' health?	Ask to see the assessments. Are they adequate, and do they identify the safeguards required to control exposure?			
-05	Are risk assessments carried out for all experiments and tests?	Ask to see the risk assessments.			
-06	Are experimental methods defined for all experiments and tests?	Observe if test/experimental methods are being followed. Do these methods cover the essential health and safety requirements?			
-07	Are waste substances disposed of in a safe and environmentally appropriate way?	Check the liquid, solid and gaseous waste receptors. How do they ensure that regulatory limits are achieved?			
-08	Are harmful gases controlled in enclosed systems or fume cupboards?	Ask how the integrity of fugitive gas control systems is assured. Look at test and inspection records for fume ventilation systems.			
-09	Is there a policy not to allow food and drink within the laboratory?	Check for signs of eating and drinking utensils. Where is the mess room?			
-10	Are antidotes available for ingestion of known toxins?	Ask to see the antidotes. Who is authorised to administer these?			
				Aspect 043 total	

ASPECT 044 Working with Visual Display Units (VDUs)

044	Audit Check	How to Verify	Act	Notes	OK
-01	Are the risks of repetitive strain injury and eye strain understood?	Discuss with computer users.			
-02	Have assessments of workstations been done?	Check if actions from assessments have been implemented to minimise the risk of eye strain or repetitive strain injury.			
-03	Are chairs for computer users adjustable in height?	Observe the chairs in use. Are the adjustments correctly set?			
-04	Is an upright posture maintained while sitting?	Observe workers' sitting position. Ask if lumbar support is comfortable.			
-05	Are VDU screens positioned to minimise glare?	Observe position and orientation of computer or other VDU screens. If facing the light, are glare filters used?			
-06	Are screens at correct height (roughly at eye level)?	Observe screens in use.			
-07	Are forearms horizontal when using keyboard?	Observe keyboard usage.			
-08	Is adequate space provided between top of legs and the table?	Discuss with computer users.			
-09	Is there a policy to take short breaks when using computers for long periods?	Discuss with computer users.			
-10	Are footrests provided when required?	Check that computer users' feet touch the ground when sitting at their workstation.			
				Aspect 044 total	

ASPECT 045 Emergency Plans

045	Audit Check	How to Verify	Act	Notes	OK
-01	Has the facility assessed all reasonably foreseeable incidents?	Ask to see what scenarios have been considered. Do these cover SHE and neighbouring facility incidents?			
-02	Is there an alarm system that can be heard by all?	Check if associates know what the alarm sounds like and when it was last tested.			
-03	Do arrangements exist for the prompt treatment of injuries?	Is there an in-house ERT or first aid/medical service? What arrangements exist to make use of local medical/hospital services?			
-04	Are procedures in place to summon assistance from external emergency services?	How are the external services called – by phone or by automated system? Have these arrangements been tested recently?			
-05	Do plans exist to mitigate the effects of environmental releases?	Does the emergency plan deal with environmental escapes and notification of the relevant authorities?			
-06	Can all personnel be accounted for in an emergency?	Ask to see the arrangements for head counts in the event of an emergency. How are contractors and visitors accounted for?			
-07	Are all people aware of what action to take in an emergency?	Ask a random sample of associates what action they would take in an emergency. Were you notified of what action to take when you arrived?			
-08	Are evacuation drills carried out?	Ask to see the records of evacuation drills and the learning actions that arose from them.			
-09	Are assembly points located in safe areas?	Visit the assembly points.			
-10	Do emergency procedures deal with terrorist threats?	Has thought been put into dealing with bomb threats, serious vandalism or terrorist action?			
				Aspect 045 total	

ASPECT 046 Use of Contracted Services

046	Audit Check	How to Verify	Act	Notes	OK
-01	Are the contractors competent to perform the service?	The fact that they have done it before is not good enough. What formal assessments have taken place to demonstrate the contractor's competence?			
-02	Does the contractor apply the same SHE standards?	Ask to see the contractor's SHE policy statement and to demonstrate its safe system of work. Check its accident/incident frequency rates.			
-03	Does the contractor have professional SHE resources?	Are these of sufficient number to be suitable for the work being carried out?			
-04	Does the contractor have on-site supervision?	Ask to meet the supervisor – *now*. Don't delay while they find someone to send to the site.			
-05	Does the contractor have suitable and sufficient tools?	Examine the condition of a random selection of contractor's tools. Especially check the condition of portable electrical tools and appliances.			
-06	Does the contractor carry out risk assessments?	Ask to see the risk assessments and work method statements. Are they adequate?			
-07	Does the contractor supply all the necessary PPE?	During the site familiarisation tours/inspections, check that contractors are wearing suitable protective equipment and that it is in a good state of repair.			
-08	Is contractor completed work checked before being signed off?	Ask what sorts of quality checks are done on contractor work. How is the company qualified to assess the work of specialist contractors?			
-09	Do contractors have suitable eating and welfare facilities?	Examine the contractor mess room and locker/washing facilities. Do they meet regulatory requirements?			
-10	Are all injuries to contractors reported?	Ask to see evidence that contractor injuries are treated in the same way as employees.			
				Aspect 046 total	

ASPECT 047 Product Stewardship

047	Audit Check	How to Verify	Act	Notes	OK
-01	Is there a written policy statement regarding product safety?	This may be integrated into the company health and safety policy.			
-02	Are associates aware of their product stewardship responsibilities?	Ask a sample of associates.			
-03	Does product design consider the full product life cycle?	Check if life cycle assessments have been done. What happens to the product at the end of its life?			
-04	Is the performance of suppliers considered before contracts are let?	Ask if Third World suppliers are used and how their SHE standards compare with those of this organisation.			
-05	Is adequate SHE information available for the product?	Ask to see MSDS, certificates of conformity and other regulatory statements to confirm the product is safe to use and be disposed of.			
-06	Is the packaging and storage of the product appropriate and safe?	View the product storage and packaging.			
-07	Is there a safe method for distributing the product?	Ask how the product is stored and transported. Is there a problem if it is stored or transported with other products?			
-08	Do emergency plans exist to handle incidents involving the product?	Ask if the public or distribution employees could be affected by some sort of distribution emergency (fire or road traffic accident).			
-09	Are customers trained in the safe handling of the product?	Ask to see evidence of training carried out.			
-10	Do product labels meet the regulatory requirements?	View a sample of the product labels.			
				Aspect 047 total	

ASPECT 048 Environmental Impact Assessments (EIAs)					
048	**Audit Check**	**How to Verify**	**Act**	**Notes**	**OK**
-01	Does the EIA contain a description of the activity/process?	Ask to be shown the most recent EIA to be carried out at this location.			
-02	Have the feasible alternatives been considered?	Check if this appears in the EIA or whether it was never documented.			
-03	Have the foreseeable environmental issues been identified?	Are *all* the foreseeable environmental issues in the report?			
-04	Have environmental pathways and receptors been identified?	Check in the EIA. Often this may be diagrammatic.			
-05	Have aesthetic measures to reduce visual impact been specified?	Check what was proposed and what consultation was done in respect to this, as it will have been a qualitative judgement.			
-06	Have technical measures to minimise environmental impact been specified?	Are the technical measures appropriate to the main environmental effects?			
-07	Has the impact of the activity on people been assessed?	Wherever possible this should be quantified. For example, the increase in noise at the site boundary.			
-08	Has the impact on other species been assessed?	Look for consideration of the effect on aquatic life, birds, bats, flora, etc.			
-09	Is there a record of applicable regulatory requirements?	Check that the regulatory record is in the EIA and not in some other database.			
-10	Does the EIA consider the effect of no action?	Ask if there could be a worse environmental consequence if nothing were to be done.			
				Aspect 048 total	

ASPECT 049 Solid Waste Disposal

049	Audit Check	How to Verify	Act	Notes	OK
-01	Is waste production managed to be the minimum possible?	What is the waste policy? What arrangements are in place to reduce waste?			
-02	Are waste streams clearly identified?	Find out what the waste streams are. Check if these are all non-hazardous. Are the waste areas on-site clearly identified?			
-03	Is waste production monitored, labelled and recorded?	Ask to see the records for the last year. Do they comply with local regulations? Is on-site waste clearly labelled?			
-04	Is the waste disposal routed to a legally approved receptor?	Check if the waste is destined for a recycling facility, waste sorting facility, landfill or incinerator.			
-05	Is all waste contained in a purpose-designed container before leaving the site?	This might include the use of skips/dumpsters or specialty refuse (garbage) vehicles etc. Compacted bales of paper/cardboard would be considered to form an integral container.			
-06	Is waste transported by an approved contractor?	Check if the hauler is aware of its legal responsibilities in relation to the transportation of waste.			
-07	Does waste undergo any form of primary treatment before leaving the site?	Possible types of treatment could include waste sorting and segregation, compaction, shredding, filtration or drying (to reduce water content).			
-08	Are waste storage areas designed to prevent the risk of ground or wind-blown pollution?	Check the storage areas. Ensure that the waste is not liable to blow away, and that the storage is protected from rain, which may wash contaminants into the ground.			
-09	Are storage areas protected against fire?	Ask how fire prevention and control relates to the storage area. Are there any sources of ignition?			
-10	Are the waste-handling companies and final destination audited?	Ask to see examples of recent audits.			
				Aspect 049 total	

ASPECT 050 Air Emissions

050	Audit Check	How to Verify	Act	Notes	OK
-01	Are the regulatory requirements understood?	Are the emission limits clearly documented and understood by those most likely to cause the emissions?			
-02	Are all emission sources clearly identified?	Ask to see the list of sources of hazardous or environmentally harmful emissions.			
-03	Are all emissions periodically quantified by measurement?	Ask to see the most recent measurements. Do they exceed the permitted levels? Is the person taking the measurements competent?			
-04	Are operating procedures designed to ensure emissions do not exceed the permit?	Review a sample of the relevant procedures.			
-05	Is emission control equipment calibrated and well maintained?	Review the calibrations. Are they sufficiently frequent? Is the equipment undamaged and fit for its purpose?			
-06	Are the emission points located to minimise the impact on neighbours?	Particularly check sensitive locations near public areas and domestic properties.			
-07	Are deviations from the permitted emissions recorded and investigated?	Ask to see the most recent examples.			
-08	Are combustion gases considered an environmental emission?	Check records relating to boiler stacks. If coal/heavy oil fired, is there an issue relating to soot blowing?			
-09	Is there a process for dealing with complaints from neighbours?	Find out how these are handled. Do they include both substantiated and unsubstantiated complaints?			
-10	Is environmental noise monitored?	What is the current noise level around the site boundary (especially at sensitive areas)?			
				Aspect 050 total	

ASPECT 051 Drainage

051	Audit Check	How to Verify	Act	Notes	OK
-01	Are foul/chemical drains segregated from surface water?	Check if surface water discharges to a water course or soak-away; that there is no risk of contamination. Are changes to the system rigorously controlled?			
-02	Is the drainage system recorded on up-to-date drawings?	View drawings.			
-03	Are the limits of wastewater composition clearly defined?	Ask to see the discharge permits. Are internal controls set lower than the permit concentrations to allow for minor errors?			
-04	Are the main drains periodically inspected?	Ask to see closed circuit television recordings of inspections.			
-05	Is the surface area draining to each gulley defined?	View drawings.			
-06	Are clayware drains protected from thermal shock?	Ask how it is ensured that high-temperature liquids cannot enter clayware drains.			
-07	Are manholes and inspection chambers periodically tested?	Check to see if manholes are numbered with flow directions marked. Ask to see the inspection records.			
-08	Are oil traps inspected and routinely emptied?	Is there a procedure requiring the emptying of oil traps? View the records.			
-09	Are redundant drains sealed off?	Look for signs of redundant plant and check if drains have been sealed off.			
-10	Is drain integrity checked by periodic pressure testing?	Ask to see the pressure test standard and results.			
				Aspect 051 total	

ASPECT 052 Soak-Aways and Ditches

052	Audit Check	How to Verify	Act	Notes	OK
-01	Is drainage to soak-aways checked for unauthorised discharges?	What checks are in place?			
-02	Are soak-aways inspected periodically for settlement?	Is there a robust periodic inspection system?			
-03	Are groundwater samples taken periodically?	Are the samples taken down hydraulic gradient from the soak-away? Is monitoring based on a risk assessment of potential contaminants?			
-04	Are plans in place to eliminate the use of soak-aways?	Look for evidence that this is included in some form of plan to which there is commitment.			
-05	Are soak-aways remote from nearby watercourses?	Check which way the groundwater flows. Does the flow increase the risk of water from the soak-away reaching the watercourse?			
-06	Are the receptors for drainage from ditches identified and controlled?	Ask if the receptors (i.e. watercourses, aquifers, etc.) are formally recorded.			
-07	Is ditch vegetation checked for signs of contamination?	Ask to see evidence that these checks are carried out. View the ditches yourself and look for signs of dead vegetation.			
-08	Are ditches secure to prevent unauthorised dumping/vandalism?	Is access controlled by fences or other security system?			
-09	Are plans in place to eliminate open ditches?	Look for evidence that this is included in some form of plan to which there is commitment.			
-10	Are potential targets periodically monitored?	Ask to see the analysis results from recent samples.			
				Aspect 052 total	

ASPECT 053 Landfills

053	Audit Check	How to Verify	Act	Notes	OK
-01	Are records of deposited materials maintained?	Ask to see the records. Do these go back to the time when the site started to use this landfill?			
-02	Are borehole records maintained?	View borehole records. Are there any problem contaminants?			
-03	Is each landfill cell fitted with vapour venting?	Check landfill for evidence of vent pipes.			
-04	Is the landfill inspected?	Ask to see the most recent regulatory inspection reports. Have actions been followed through?			
-05	Is newly deposited waste covered daily?	Check for evidence of exposed material.			
-06	Is the landfill compacted?	Are the spreading machines designed to compact as well as spread the waste?			
-07	Is the landfill lined with an impervious material?	Ask to see cross-sectional drawings of the landfill. Look for evidence of a clay or other impervious lining.			
-08	Is leachate from the landfill regularly monitored?	Where is the leachate monitored? What are the most recent results? Where does the leachate discharge?			
-09	Are the landfills subject to independent audit?	Ask to see examples of third-party audits.			
-10	Are measures taken to control wind-blown debris and exclude the public?	Is the landfill secure and fenced to prevent wind-blown debris carrying beyond the boundary?			
				Aspect 053 total	

ASPECT 054 Storage Tank Secondary Containment

054	Audit Check	How to Verify	Act	Notes	OK
-01	Are tank bunds/dykes periodically inspected?	Ask to see inspection record.			
-02	Have tank bunds/dykes had a recent water containment check?	Ask to see the test report.			
-03	Are bund/dyke drain valves kept routinely locked closed?	Check during the plant familiarisation tour if bund/dyke valves are opened only for rainwater draining. Is the rainwater analysed before it is drained?			
-04	Is all equipment requiring regular operator intervention outside the bund/dyke?				
-05	Is the weather seal between the tank base and the tank intact?	Ensure that weather seals are intact to prevent rainwater corrosion of the base.			
-06	Do all overflow points discharge inside the bund/dyke?	Look for all points where overflows discharge and ensure they are within the secondary containment area.			
-07	Is the bund/dyke capacity suitable for the tanks contained in it?	Rules vary from country to country, but the capacity should be no less than the volume of the largest tank.			
-08	Is the bund/dyke capacity suitable to contain contaminated firewater?	Particularly check this if there are installed sprinkler/water quench systems.			
-09	Do leakage test probes exist for single-walled underground tanks?	Work from the principle that underground tanks will probably be leaking. How do they check for this?			
-10	Are redundant tanks removed?	Check during the plant familiarisation tour.			
				Aspect 054 total	

ASPECT 055 Drum Storage

055	Audit Check	How to Verify	Act	Notes	OK
-01	Are drums always stored on bunded pallets?	Do a visual check.			
-02	Are drum storage areas designed to drain spillages to a containment area?	Look for evidence of curbing and a collection sump. Where does the sump discharge?			
-03	Are records maintained of all materials stored in drums?	Ask to see the historical records.			
-04	Are all drums clearly labelled (including intermediates and wastes)?	Do a visual check. Especially look for wastes and intermediates that may not have proprietary labels.			
-05	Are drums protected from the weather (rusting or ultraviolet attack)?	Do a visual check. Metal drums suffer from rusting and plastic drums from ultraviolet light embrittlement.			
-06	Are all drums in a good state of repair?	Look for signs of impact damage or missing filler caps.			
-07	Is there a properly designed method for removing the drum contents?	Do a visual check. Look for examples of drum pumps or tilting pallets. Check in both cases how minor spills are handled.			
-08	Are drums protected from extremes of temperature?	Do a visual check for barrelling of the drums (overpressure) or sucking in (vacuum).			
-09	Are empty drums recycled?	Check if drums go back to the supplier for reuse. Recycling as rubbish bins does not count.			
-10	Has there been a fire risk assessment?	Have the contamination consequences of a fire in the drum area been taken into account?			
				Aspect 055 total	

ASPECT 056 Liquid Loading/Unloading

056	Audit Check	How to Verify	Act	Notes	OK
-01	Are transfer points paved and curbed?	Do a visual check.			
-02	Is the containment surface suitable?	Tar macadam is inappropriate for hydrocarbon material, concrete is unsuitable for acids, resin surfaces are unsuitable for plastic pellets, etc.			
-03	Are construction or subsidence joints effectively sealed?	Do a visual check on the condition of caulking. Look for examples of subsidence cracking.			
-04	Can the containment system contain the full contents being discharged?	This is especially important if the discharging container is a large road or rail tanker.			
-05	In the event of a spillage can the product flow be isolated easily?	Check if this isolation can be done remotely. If there is a discharge/spillage, will the isolation point be affected, preventing its operation?			
-06	Are the contents of hoses drained to a safe place?	Check what happens to the contents of the loading hose once loading is completed.			
-07	Is a system in place to prevent vehicles driving away while coupled?	Ask how they prevent drive-aways when loading/unloading hoses are still coupled?			
-08	Are vehicles earthed during the loading/offloading process?	Do a visual check.			
-09	Is spillage from sample points collected ?	Observe sample points in use.			
-10	Do procedures exist to control the washing out of containers?	Ask if container or tanker cleaning is permitted, and if so what happens to the washings.			
				Aspect 056 total	

ASPECT 057 Groundwater Abstraction

057	Audit Check	How to Verify	Act	Notes	OK
-01	Is there a regulatory permit for the abstraction?	Ask to see the permit.			
-02	Is the pumping level maintained constant in normal use?	Look out for evidence that the pumping levels are dropping continuously.			
-03	Are samples routinely taken from the well?	Ask to see sample analysis results. Are the analyses appropriate to what could be anticipated?			
-04	Are the well draw-down areas protected to prevent contamination?	Ensure that well head areas are secure or enclosed.			
-05	Is the zone of influence of the well known?	Check if the well could be abstracting water from a neighbour's property.			
-06	Is the well string pump a good fit in the well?	Ask if internal recycling within and around the well pump could be leading to excessive power consumption.			
-07	Is the well liner inspected periodically?	When was the last inspection?			
-08	If used for drinking water, is the water treated?	Ask what sort of treatment system is used. How does this fit with the regulatory requirements?			
-09	Is well head pipework protected against temperature extremes?	Consider whether frost damage or thermal expansion could be a problem when the well is shut down.			
-10	Was a historical review conducted before the well was installed?	Did the historical review indicate if toxic or ecotoxic chemicals could have contaminated the well catchment area?			
				Aspect 057 total	

ASPECT 058 Ground Contamination: Historical Review

058	Audit Check	How to Verify	Act	Notes	OK
-01	Are all the current contamination sources known and recorded?	Ask to see the historical review report.			
-02	Has a review been done of the possible historic uses of the land?	Ask to see the historical review report.			
-03	Is the hydrogeology of the site known?	Ask to see a copy of the hydrogeological map.			
-04	Has an inspection been done to find evidence of contamination?	This inspection should have looked for evidence of old spillages, leachate or existing of old tipping sites.			
-05	Have all historic contaminants been identified?	Ask to see the list.			
-06	Has a literature review been carried out?	Look for evidence that they have reviewed company records, local authority and regulator records, aerial photographs, retired employees, etc.			
-07	Has it been established that the site is not of special ecological interest?	Ask to see flora/fauna survey reports.			
-08	Has it been established that the site is not of archaeological interest?	Ask to see detailed local maps.			
-09	Has the historical review been used to define the sampling strategy?	Ask to see a demonstration of the link between the sampling strategy and the historical review findings (often found in a consultant's report).			
-10	Have all environmental targets been identified?	These might be watercourses, ground or ground contamination, potable water abstraction, etc.			
				Aspect 058 total	

ASPECT 059 Site Investigations

059	Audit Check	How to Verify	Act	Notes	OK
-01	Has soil sampling been carried out?	Ask to see trial pit or gas probe survey results.			
-02	Was the sampling strategy based on a historical review?	Ask to see the historical review summary.			
-03	Did a competent person carry out sampling?	Were samples collected and stored in a way that would not change the contamination levels? Did samples follow a chain of custody to the laboratory?			
-04	Were samples representative of the area being studied?				
-05	Was sample analysis carried out in a laboratory certified by a suitable authority?	Check if laboratory is registered with the EPA (United States), NAMAS (Europe) or other equivalent body.			
-06	Are monitoring boreholes sited on the downstream boundary?	There may be other monitoring boreholes, but the down hydraulic gradient may indicate contaminants migrating from the site.			
-07	Does borehole design prevent pollution ingress into hole?	Check for the presence of an aboveground upstand to prevent contaminants leaking down to the groundwater. Are the boreholes locked?			
-08	Are unused boreholes properly sealed?	Check state of boreholes that are not regularly monitored.			
-09	Are boreholes designed to monitor water at a specific depth?	Check the piezometer tube design to see if the slots are set at the required depth.			
-10	Is there a clearly specified borehole sampling method?	Ask to see the method statement.			
				Aspect 059 total	

ASPECT 060 Waste Minimisation

060	Audit Check	How to Verify	Act	Notes	OK
-01	Is there a policy of waste minimisation?	Ask to see the records of waste produced over the last 5 years. Is it reducing?			
-02	Is there a waste segregation process?	Ask to see what is segregated and how it is controlled.			
-03	Is waste recycled as much as possible?	Ask to see examples of waste recycling. Is everyone involved?			
-04	Are the offices involved in waste reduction?	Is paper usage minimised and paper recycled? Are other consumables (ink cartridges, cups, etc.) recycled?			
-05	Is the use of recycled materials part of the purchasing policy?	Ask to be shown examples of purchases involving recycled materials.			
-06	Are packaging materials all recycled?	Check if suppliers' packaging is returned to source and if maintenance packaging (cable reels etc.) is returned.			
-07	Is production waste measured and controlled?	Is production efficiency as high as possible to minimise waste?			
-08	Are surplus materials disposed of for gainful use?	Are they sold or given to schools/charities?			
-09	Is old IT equipment sent for recycling or reuse?	What happens to obsolete computers?			
-10	Are alternative uses considered for waste materials?				
				Aspect 060 total	

Index